Mating

About the Authors

The three authors previously co-authored the best-selling book FINDING YOUR BEST PLACE TO LIVE IN AMERICA, which is revised biannually.

M. Ronald Mingé, Ph.D., is a clinical psychologist and professor at C. W. Post College of Long Island University. His experience as a marriage and family therapist has extended over twenty years, and he has published extensively. Dr. Mingé is married and the father of four daughters.

George A. Giuliani, Ph.D., was the Dean of the Undergraduate School of Education at C. W. Post College for many years and is presently a professor and clinical psychologist with a practice in marriage and family therapy. He is the director of a consulting firm and has published extensively. Dr. Giuliani is married and the father of two sons and a daughter.

Thomas F. Bowman, Ed.D, is a professor of educational administration at C. W. Post College of Long Island University. He was the Dean of the Graduate School of Education for many years and has done post-doctoral study in marriage and family therapy. He is familiar to many from more than 300 appearances on radio and television as well as feature articles in newspapers and magazines. He is married and the father of two daughters and a son.

Mating

by
M. Ronald Mingé, Ph.D.
George A. Giuliani, Ph.D.
Thomas F. Bowman, Ed.D.

RED LION BOOKS · NEW YORK

1st Printing-November 1982

Mating

Designed by: Morry M. Gropper

Composition: A&S Graphics, Inc.

Library of Congress Catalog Card Number 82-62175
ISBN: 0-940162-01-6
Printed in the United States of America

Dedication

Dedicated to the beloved consequences of our own Dynamic Mating:

George

Roger

Kathy

Diana

Claudia

Sharon

Kristen

Marisa

Jimmy

Jeanine

Contents

Contents

Acknowledgments

The tireless efforts of Mark McCoy as research director for this project are sincerely appreciated. As usual, he accomplished wonders.

Marilyn Oser, Ph.D., Mary Lou Kallman, Judith Kroe-Menton, Carol Hill, and Myra Verasco all contributed creative research and unflagging enthusiasm throughout this major endeavor, and they made research both productive and a pleasure.

Joyce Grinnell, Marie Bowman, Marilyn Mingé, and Carol Giuliani aided in the development of the MATING principles and devoted countless hours to manuscript preparation.

Carol Dioguardi and Kathleen Ross provided supportive services with never failing charm and effectiveness.

The insights and assistance of our excellent attorney, Bruce Gold, are always appreciated.

Patrick Shanahan, Ph.D., was generous once again with his time, expertise, and inspiration.

Gerald Albert, Ed.D., contributed excellent suggestions and reactions, and his early work in the field of mate selection has been most valuable to our research.

The consideration of Fran Riordan is greatly appreciated.

A number of splendid librarians have aided in this

project. Special assistance was provided by the following:

Wendy Roberts, Iris Irwin, and Marilyn Rosenthal, as well as the Periodical Department staff, C.W. Post College Library.

Antoinette McGrath, Head Reference Librarian, and June Schwartz, Periodical Librarian, Huntington Public Library, Huntington, N.Y.

To all of the above, our heartfelt thanks.

The ideas and research evidence presented in MATING are likely to be shocking, fascinating, and revolutionary.

Why?

Because it presents startling new information and a scientifically based system for finding excellent mates and building loving, successful mating relationships. And this long overdue approach is dramatically different from what has been going on for centuries.

We humans have been mating for four million years, give or take a month or two, and with all that practice you might think we'd be getting it pretty well perfected by now.

Well, we've moved out of the Stone Age, but most of our mating practices are positively medieval. And traditional mating practices have, without question, produced results which are catastrophic.

Divorce rates have skyrocketed to the point where *one-half* of recent marriages are now projected to end in divorce. These are the same odds as the flip of a coin!

We have been marriage therapists and researchers for many years, and we know that these divorce rates are completely absurd and totally unjustified. These are not just mindboggling statistics, they represent frightened and badly injured individuals, adults and children alike.

And we all know that millions of other couples stay together even though they are miserable and unhappy with each other.

Our effort in MATING has been to bring to you excellent research and clinical evidence which has been accumulating over the past decade or so. Most of it has remained hidden from the public in scientific journals and behind walls of highly specialized language.

But people have been snarled in myths and weighted down by misconceptions for centuries. They are often afraid of new information if it doesn't fit what they've been taught to believe. Many don't seem to want anyone to look at love and marriage success too carefully, apparently fearing that they are too fragile or mysterious to be understood.

"Better to leave it natural," they exclaim. "Leave it a mystery."

Perhaps we could if people were enjoying mating success, but not with 50 percent failure rates!

We have made every effort to bring you the research and clinical findings with familiar language and practical applications. We've thrown in some humor, because even though it's a serious task we've taken on, mating can be funny too.

We have wrapped the research findings, supplemented by our experience, in a complete and cohesive new system for making your best mating decisions. The Dynamic Mating System has been developed over several years and reflects our interest in combining clinical concern with research objectivity.

Finally, you may be pleased to see that we have treated love as a vital part of the Dynamic Mating System and have provided many strategies for building truly loving and successful relationships for a lifetime.

1

Your Mating Decisions

WHETHER YOU ARE LOOKING for a mate or are already married, you still have some critically important decisions to be made. They are, in fact, the most important choices of your life.

Who you live with, and how you live together, will have an incredible impact on your happiness, personality, physical and mental health, success, leisure time, savings account . . . just about every facet of your existence.

And your mating decisions will have their effect for generations, determining the well being of your children and your children's children.

Mating should be, and can be, a joyful, mutually productive, loving, lifelong process. It is more than making love. It is more than marriage. But making it happen requires the use of your mind as well as your heart.

How well are people doing with their mating decisions?

We are a long way from the kinds of mating relationships we could be enjoying. Recent newspaper headlines are a devastating commentary on what is happening to people as a result of today's mating practices: [1,2,3,4]

50 PERCENT OF RECENT MARRIAGES
WILL END IN DIVORCE COURTS

ONE HALF OF NATION'S KIDS
TO LIVE IN BROKEN HOMES

SINGLE-PARENT FAMILIES
DOUBLE IN LAST DECADE

DIVORCE RATES THREE TIMES
AS HIGH AS 20 YEARS AGO!

People have had so many problems that today's marriages should be classified as a national disaster of insane proportions!

And it is growing worse every day.

You know the anguish behind the headlines is too great to ignore. You've heard others describe their anxiety about marriage and the problems it can bring. You may have experienced much of that concern too.

Whether they are looking for a mate or are already married, people are worried, and justifiably so. No one wants to become a divorce statistic. No one wants to subject children to the pain and confusion of being separated from a parent.

In our own practices of marriage therapy, we have listened for thousands of hours to questions like these:

- Why can't I find a good husband (or wife)?
- What is going wrong with my marriage?
- What kind of person should I marry?
- Why are all my friends getting divorces?
- What's wrong with our society?
- Should I get married again? Who would marry me?

And we have heard all kinds of stories about marital problems and what they do to adults and children alike. The difficulties faced by this young woman are not unusual in the therapy office:

> Carlene was visibly upset. Her hands shook as she lit another cigarette off the first, and she trembled on the verge of hysteria.
>
> "I just don't understand it! We seemed so right for each other," she said. "We both love adventure. On our honeymoon we went to Las Vegas and Hawaii both. Our sex together has always been terrific. Even after he's hurt me so much, I'd still go to bed with that rotten . . . I'd better watch my language."
>
> She was talking about her husband of three years, who had just moved out of their home.
>
> "Why didn't he love me? It's just so unfair! I did everything for him, but he wouldn't tell me he loved me . . . except during sex. He'd just complain that I was getting too fat, or that he didn't like people hanging all over him."

Up, and Up, and . . .

Proportion of Marriages Begun That Ended or Will End in Divorce[1a]

Year	Percentage
1870	8%
1890	10%
1910	14%
1930	23%
1950	30%
1970	48%
1980	50%

Would you like to make any guesses for the year 2000?

Carlene's face twisted with anguish as she thought of the humiliation she had endured for so long and apparently for so little.

"Maybe the difference in our families was too great. Nobody shows any affection in his, and in mine everybody is always kissing and hugging. He was always gone, off with those stupid buddies of his, playing baseball or going hunting or something. Sometimes I wondered if he was, you know, more interested in guys.

"And then, when I found the lipstick on his underwear, and he finally admitted he had a girlfriend . . . what went wrong? Why couldn't he love me?"

And she wept for a long time.

The Destruction From Bad Marriages

You've seen some of the terrible consequences of contemporary mating attitudes, probably close up. You have watched people you know and love as they struggled to cope with a bad marriage, trying to salvage a bit of pride, hurt and bewildered and fighting back, as they drifted closer and closer to the brink of divorce.

If you have been there yourself, no one will convince you that such problems are other than the most hurtful and exhausting of your life.

All too often the brutal effects on the children, victimized by circumstances beyond their control, are too much to witness. Torn between two parents they love, powerless to intercede, they usually are sure that they somehow caused the family's misery.[5,6]

Research shows that large numbers of children are scarred permanently by the divorce experience.[7] Racked by guilt and depression, they will be severely handicapped in their own mating efforts later in life, and in their turn are likely to pass problems along to their own children.[8,9]

The cost of mating errors is appallingly high for individuals and for their families, and the hidden cost to the nation itself is astronomical.[10,11] The drain of time, energy, and finances caused by marital problems is enormous, because people who are caught up in the emotional wringer of mismating cannot be as effective or productive as when they are happy and enthusiastic about their lives.

Shocked and alarmed, sensitive people are becoming more and more aware of how vitally important their mating decisions really are, and how little they have known about finding a mate who is truly right for them.

The Dynamic Mating System

What has been needed is a system for answering the critically important questions about who to marry, and how to build a successful and loving relationship for a lifetime.

To accomplish this, a radically new view of the mating process is needed. Why? Because traditional attitudes have failed miserably, and recent changes have been no better. Whatever is going to work has to be a radical departure.

The process we have developed, called the Dynamic Mating System, is intended to provide strategies for finding an excellent mate and building a lifetime of productive love and companionship.

It is not intended to be controversial, but it is forced into that posture. For one thing, it replaces old myths and recent attitudes about mating, and both of these points of view have their staunch supporters and true believers. For another, legions of self-styled "romantics," brainwashed by song lyrics and conditioned by fiction, are always ready to attack any effort to demystify love, even with their full knowledge of the disastrous state of contemporary mating in America.

"When it comes to romance," they stoutly insist, "ignorance is bliss."

The Dynamic Mating System is controversial because it does not fit in with the "new chic" of today's emphasis on temporary and superficial relationships. We cannot condone, for example, the destructive advice of popular writers and columnists who advocate extramarital affairs as "beneficial" to a marriage. The research findings too strongly indicate otherwise.[12,13,14,15,16]

Listen to the attitudes expressed by a great many "modern" men and women as they discuss the relationships in which they are involved. Here is an extremely well-groomed man in his middle-thirties:

> "Listen, you can't tell me, or any of the guys I know, anything about women and what they want. We've got it down pat. I personally have been going to singles bars for fifteen years, and I know every line. You can't teach me or my friends anything new. We already score every time."

From our point of view, based on extensive research and many years of clinical experience, a change in mating attitudes is not only in order, it is imperative.

Men and women are fed up with unhappy marriages,

If It Works, Use It

It is said that 4000 years ago women already had developed an effective vaginal contraceptive. They used a suppository with an alkaline base, and human sperm require an acid base for functioning.

What was their magic ingredient?

Crocodile dung. *Not* available at your local pharmacy.

loneliness, the meat racks of singles bars, alimony, and the struggle of raising their children alone.

They don't want or need any more lame rationalizations that divorce is "...just a natural process," or "You can change it into a creative experience if you really try." This kind of sanctimonious justification of terrible events follows all kinds of disasters, but it doesn't make them any less painful.

The Need for Answers

What people need are answers. They need to know *WHO*, and *WHETHER*, and *WHAT TO WATCH OUT FOR*. They also need to know how to make relationships *WORK* for continued growth and satisfaction.

MATING has been written so you can find personal answers to such compelling questions as:

- What kind of mate am I?
- What kind of mate is best for me?
- How will I attract such a person?
- Where can I find such people?
- Which people are bad risks for me?
- What kind of person is my lover or spouse?
- What causes divorce?
- Should I marry (or remarry) now, or later, or ever?
- How can I increase my chances for a loving relationship which will last for a lifetime?

Accurate and practical answers to these kinds of questions have not been easy to come by, either in the past or in the present. People have had to wrestle with the issues by themselves, and they have had to make their best guesses based on few facts and a superabundance of misconceptions.

They have had to rely on the advice of friends and relatives, who sometimes were helpful but more often were not, because they were just as handicapped themselves by wrong information.

People have felt overwhelmed by the task of weighing the pro's and con's of their mating decisions. As one woman told us:

> "I tried to think of everything, but after a while I just got more and more confused. Finally, I decided 'Enough is enough' and I married Elliot. I'm still not sure why, though. I know I liked his car and his parents. Probably in that order."

But the stakes are too high and the consequences too vital to leave mating decisions to a cross-your-fingers-and-hope-for-the-best approach. It is dangerous to base mating choices on bits and pieces of unrelated information or charming but unreliable folklore.

The Purpose of the Dynamic Mating System

The Dynamic Mating System, which is the heart of MATING, has been specifically developed for two purposes:

(1) It provides a straightforward and comprehensive system for finding a mate who is right for you, or for evaluating the mate you already have.

(2) It offers specific strategies and procedures for building a dynamic mating relationship, or for understanding and improving your present marriage.

Because it is a systematic process, you are much less likely to overlook essential information or forget critically important issues.

We have designed special new tests for the Dynamic

Mating System to answer your specific questions and evaluate major areas of concern. As we said, we want you to make your decisions with *both* your heart and mind.

Does our emphasis on intelligent decision making mean that we are somehow opposed to romantic and passionate love? By no means. As you will find throughout the book, love in all its permutations is an essential part of the mating process.

But as long as we are going to fall head-over-heels in love, crazy about someone's wit and sexy earlobes, with our glands pumping out secretions for all they are worth, we might just as well increase our chances of falling for someone who will be absolutely wonderful in other ways too.

When the mating is right, when two compatible and loving people find each other, the results are excellent.

The results are even better when they thoughtfully set out to improve their relationship, with plans and strategies and mutual objectives.

And when they make a genuine effort to take care of each other, to communicate their feelings and to solve problems together, the consequences are nothing short of spectacular!

This kind of relationship is what MATING is all about.

Looking for a Mate?

The U. S. Bureau of the Census says there are 53,000,000 unmarried adults in the nation. Twenty-three million are men and thirty million are women.

Let's see. Thirty million divided by twenty-three million . . . does anyone have a calculator?

2

Cinderella and The Fly

IT WAS LOVE AT FIRST SIGHT . . .

. . . and they lived happily ever after.

Myths about mating have been the stuff of story and song since time immemorial. Their themes have been passed along from generation to generation, and the fascination with love gained and love lost no doubt will endure forever.

Some are beautiful classics, capable of bringing us joy or tears each time we hear them. Others give us temporary entertainment and are quickly forgotten. In either case, however, their themes linger on, influencing and molding beliefs and attitudes about romance and the way it is supposed to be.

The problem is that the underlying messages are, for the most part, misleading, inaccurate, and potentially dangerous.

They are myths, and not to be trusted.

It's not that people believe every line or try to imitate every deed. What would happen, for example, if you did what they do in the popular romantic novels sold by the millions next to the laundry detergents and canned corn?

A lusty man of action, usually a pirate rebel chief named Dirk or Colin, can be counted on to kidnap the brave, fiery, and unaccountably demure beauty who is almost always of royal blood, preferably English, and named Rebecca. In no time at all, because he is a *very* lusty man of action, he will attempt to tame her with the "raw power of his unbridled lust."

Try that at your next church social!

Have you noticed how he will apologize soon after the deed, but never during? He'll mumble something about how he couldn't help himself, and besides, he thought she was someone else.

Rebecca has mixed feelings by this time. It was, after all, an exciting date. On the other hand, she *is* embar-

---

From Myth to Mrs.

Some common myths for brides:

- A spider found crawling on the wedding dress means good luck.
- Dreaming of your wedding day means good luck.
- It is bad luck to look in the mirror after you are completely dressed in your wedding clothes.
- It is good luck to have your hair and veil arranged by a happily married woman.
- To put your bare feet on the floor on your wedding night is bad luck.
- If you break something on your wedding day, it's a sure sign that you will quarrel with your mother-in-law . . . and your husband will take her side!

rassed by her enthusiasm so soon after the excitement started, so she decides to denounce him and proclaims, "I shall be avenged, Dirk or Colin!"

Of course, he was the cleanest pirate rebel she had met, and considering the terrible adventures she's already endured, he wasn't such a bad guy. His smile was nice, and he was so masterful!

You just know these two kids are going to get together by page 256!

You do not have to be Betty Friedan to recognize the myths perpetuated by so many of these novels. Active man, passive woman. Fate throws them together. Their loins meet before their parents do. She's angry and hurt, but he rescues her from Ghengis Khan or an evil slave trader, or some other fate worse than death. Love conquers all.

We have all heard these themes countless times, and they have had an irresistible effect. Even when we listened at our mothers' knees, the stories sounded something like this to our little ears:

> "Careful, Beauty. The world is a threatening place. It's best if we keep you locked up here in virginal isolation.
>
> "Ooops! Careless, clumsy Beauty. Can't even sew. Jabbed your little thumb, and now you'll have to sleep passively forever. Unless, that is, you get rescued through blind, dumb luck.
>
> "Oh, swell! Here comes the handsome prince. He's a confident guy who knows how to make things happen. He'll give you a big kiss on your pallid lips. Aha! Now *that* is how to arouse a woman!
>
> "Climb up on his horse, Beauty. No, no, sit in the *back*. Now remember, meals on time, no headache excuses, and don't let the kids do hard drugs. And you know what, Sweetie? If you're lucky, just maybe he'll keep you!"

This tale not only provides an example of how passivity is ascribed to women and effectiveness to men, it also contains the most persistent mating myth of all. It is usually boiled down into such phrases as:

"Whatever will be, will be,"
"Que sera, sera,"
"We are powerless in the face of love,"
"Good things come to those who wait," and
"Fate will bring us together."

We refer to it as the Love Lottery. Can't you just hear the pitchman now?

"Step right up, folks, and take a number. If you wait long enough, your mate might turn up. If you don't wait, you can't win. What's that? You've been waiting for nine years? Okay, lady, here's a consolation prize. Take him or leave him. Take a number and wait, folks!"

This approach is wonderfully romantic, because it implies that your true love is wandering about out there somewhere, and fate alone will bring you together when the time is right.

You might be washing your underwear, or sinking

Those Dirty Books!

So severe were Victorian ideas about segregation of the sexes that *Lady Gough's Book of Etiquette* included rules for proper people's bookshelves. It forbade the placement of books by male authors next to those by female authors, unless, of course, the authors happened to be married to each other.

Well, you never can trust those books! We've seen lots of them trying to sneak under each other's covers late at night.

into a swamp, or doing just about anything, when suddenly out of nowhere, your sweetheart will materialize.

You'll know each other instantly, and you'll fall helplessly, hopelessly in love. You can only hope that your true love won't be 106 years old and already married.

Statistical probability tells us that funny quirks of fate sometimes will bring extremely well-suited people together. But what are the odds of meeting the love-of-your-life in the way we were told this couple met?

THE FLY

It was a gorgeous summer afternoon, and the bus was packed with passengers. They were both loaded down with large bundles, and they couldn't help but notice each other—they were squeezed together in a most pleasant way.

She was lovely in a blue chiffon dress, and he felt very elegant in his blazer and grey flannel slacks. Elegant, that is, until he realized that the zipper on his fly was open. Perhaps he felt a draft.

At any rate, grateful for the crush of the crowd, he hastily zipped as unobstrusively as he could, and then he smiled at the young woman.

"Sure is crowded," he observed.

"It certainly is," she responded brightly, smiling her warmest smile. But alas. The bus had arrived at her stop. She turned to go.

Then, to his shocked amazement, he felt a vigorous pulling at the front of his pants!

And she turned back in confusion, for she had felt a very strong tug at the skirt of her dress!

They both looked down.

The zipper had closed on her blue chiffon dress!

He stammered something in dismay, and her

mouth fell open in total bewilderment. They both began to blush furiously at what the other passengers must be thinking.

Saying something like, "Gee, I wonder how that happened," he grabbed the front of his pants with both hands and tore at his offending fly.

Nothing happened.

He strained, face red and grinning fiercely, unbelieving at his luck, pulling with all his strength. Everyone was staring.

The zipper refused to budge.

She was tempted to try the stupid zipper herself, but the image was too much.

They traveled together like that for one more stop, locked in ludicrous exertion, and then he gallantly offered to get off the bus with her. After all, how could she refuse?

They stood there, very close together, as the bus pulled away, its passengers gaping back in disgust. Some people!

It didn't take her long to invite him to stumble along with her to her parents' home. They held their packages low in front, so the neighbors wouldn't know.

We were never told what was said when he was introduced to her parents at the front door. And we never learned which they finally decided to take off first, the dress or the pants.

But as you might guess, they lived happily ever after.

As happy as we are for this couple, we cannot encourage you to depend on the Love Lottery to bring you your best mate. The phrase, "If it's meant to be. . ." is a rationalization, used primarily by people who are afraid to make an effort or are unwilling to gather enough information to make intelligent choices.

The startling fact is that the majority of the people in our society live with a kind of "wait and see" attitude. The consequences can be disastrous. Why rely on it, when the right mate for you can make your life wonderful, and the wrong person can be entirely destructive.

Doing What Comes Naturally

Another myth which has misinformed millions is that the mating process is supposed to be "all natural," like yogurt and sunflower seeds.

After all, Bambi and his girlfriend knew how to do it, the jellyfish couple bobbing in the ocean doesn't have any trouble, and even the little amoeba, if it gets lonely and desperate enough, can split itself in two, although undoubtedly with the greatest reluctance.

Surely humans should be able to go through the mating process without bumping heads or forgetting to introduce themselves.

This view is largely an outgrowth of instinct theory. The problem with it is that the higher the species is on the phylogenetic scale, the more complicated its behavior is, and the more important learning becomes. This is true even for mating.

As if it weren't involved enough already, the expectations of society add another complicating dimension. As a result, human mating depends on intelligent problem solving, and has relatively little in common with the amorous adventures of chickens and clams.

Ignorance is Bliss

The "doing what comes naturally" myth is related to another often fiercely held belief that love and science cannot be mixed.

Many seem to feel that when it comes to romance, ignorance increases the chance of bliss. The following is a fairly typical statement:

> "I believe that 200 million other Americans want to leave some things in life a mystery and right at the top of things we don't want to know is why a man falls in love with a woman and vice versa. . ."[1]

What makes the statement unusual is that it was made by a United States Senator, Senator Proxmire of Wisconsin, railing against federal funding for research on love and mating practices.

As much as we enjoy a good mystery too, we disagree wholeheartedly with Senator Proxmire and others who feel that love is too sacred to be studied. The costs of mismating are staggering. The pain of mismating can be overwhelming. The effects of mismating can last for generations.

Nothing deserves immediate research attention more than does the question of how to help people find happiness and success through selecting the right mate and then growing with that person for a lifetime of fulfillment.

From a congressional point of view, Senator Proxmire and others might consider the billions of dollars lost each year because of mating mistakes. If we could improve by even five percent the quality of mating in America, it surely would rank among the most important developments of our generations.

The indoctrination that "love and science cannot mix" is encouraged by our English language, because most of the words dealing with planning, thoughtful analysis, and research have somehow become associated with coldness, or even worse, evil intent.

"Premeditated" sounds like murder.

"Calculated" goes with cruelty and iciness.

Such words as "deliberately," "purposely," and "intentionally" are heavy weapons in the arsenal of a prosecuting attorney.

It is as if it is unfair or un-American to plan, organize or solve problems in an intelligent and systematic manner.

Our concern about the anti-scientific mating myth does *not* mean that we believe the concept of love is going to be reduced to mathematical equations or the measurement of glandular output. The reason is that love is a lot like electricity. No one really knows what it is, despite all the research. However, careful and systematic studies and practical applications of findings have produced some wonderful ways to use electricity to benefit mankind. The same can happen for love.

We are going to ask you to suspend any anti-scientific, anti-systematic biases you might have. Put them on "hold." Yes, it can be difficult—as difficult as entertaining the possibility that your mother's apple pies might have been only second-rate. But try. The results will be worth it.

Clearly, the dusty old myths about mating have caused confusion and misery for too long. Uncounted millions of people have been sabotaged by them, never to know the joys and triumphs of finding and growing with the mates who would have been best for them.

Modern Myths

Another source of misconception and difficulty is the recent and dramatic revolution in attitudes toward sex and relationships. These changes have produced a set of modern myths which have had a significant impact on mating practices.

The modern myths comprise a new and contradictory set of expectations about mating. The new rules have had a dramatic effect on our nation's feelings about love and marriage.

Permissiveness has become a watchword, and the so-called sexual revolution has created totally new uncertainties.

Only a few years ago, or so it seems, young people were told about not kissing on the first date, the importance of being respectable, how to block "Roman" hands with quick elbows, and what cold showers were for.

All right, so not everybody took it seriously, but at least they were *told*.

Today, the sexual escapades and romping infidelity of famous personalities fill the pages of magazines and newspapers. The romantic novels get steamier and steamier, as the hands on the illustrated bookcovers slip lower and lower. Advertisements for birth control

---

A Divorce Rate Sampler

Divorce Rates in Selected Countries[2a]

Country	Number of Divorced Persons Per 100 Married People
United States	11.28
Sweden	9.69
Switzerland	7.58
Australia	5.41
West Germany	4.91
Czechoslovakia	4.39
France	4.09
Scotland	2.45
Japan	2.33
Israel	2.08
Phillipines	.98
Spain	.54
San Marino	.04

products are displayed prominently, with pictures of radiant couples cavorting through springtime woods. Record numbers of men, women, boys, and girls rush home daily to watch the promiscuous adventures on the daily "soaps."

The changes in attitudes have had an effect all through society. Adults have to think twice about going to a movie, even one rated PG, with their own parents or in-laws. What do you say after watching "superstars" demonstrate "superorgasms?" Do you ask, "Did you catch that last multiple, Mom?"

Little boys and girls are now haunted by the dreaded possibility of being the last virgin in Monroe Junior High School. And not a few of their teachers harbor the same fear!

We have come to expect that the marriage counselor will be separated or divorced. It's hard to tell if people are joking when they advise, "Why don't you settle down and get married a few times." About newlyweds they say, "Well, if it doesn't work out they can always get a divorce."

Genital herpes now afflicts 20 million Americans, with half a million new cases annually.[2] A frequent question for advice columnists is:

> "How do I ask, without appearing offensive, whether my partner for the night is infected? Do I simply say, 'I am assuming you don't have herpes. Am I right?' Or do I finesse the question by asking, 'Is there anything I should know?'"

Not to worry, as they say in New York City, for in the "Big Apple" a new exclusive dating service has emerged. Exclusive, that is, for herpes sufferers. At least it provides the peace of mind that members are not spreading the misery.[3]

"Living together" has become such common practice that in the past ten years the incidence of POSSLQ has increased by more than 200 percent.[4] No, it's not another venereal disease. It is how the Census Bureau makes an acronym out of "Persons of the Opposite Sex Sharing Living Quarters."

A recent study in Sweden indicates that 99 percent of all couples there live together before marriage and one out of three children is born out of wedlock.[5] Does living together before marriage lead to better marriages later? It hardly seems so, when Sweden has the eighth highest divorce rate in the world.[6]

The problem with "living together" arrangements is two-fold:

(1) Couples bring all of the same problems to live-in arrangements that they bring to marriage.
(2) Reduced levels of commitment to the relationship mean too little effort usually is made to overcome the mating difficulties which arise.

The key to solving mating problems is not to find easier ways to end relationships. It is to develop better ways to make relationships happy and productive.

Probably the greatest damage from the so-called sexual revolution, however, has been its influence in "trivializing" relationships and replacing marriages with a jar of Vaseline. In our practices of psychotherapy we've encountered large numbers of men and women

---

Not Much Support

According to a Census Bureau survey done in 1979, the majority of divorced fathers do not make child support payments. Only 43 percent of the mothers reported they received payments, and payments averaged $1900 per year.

No comment.

who consciously avoid becoming emotionally involved with anyone. As one thirty-year-old client described it:

> "I'd rather meet and enjoy as many lovers as possible, maybe three or four different ones each month. I'd even settle for one or two. All my friends have been divorced, and I'm just not going to go through the hell they've had to put up with! I simply do *not* think it is worth it!"

We are not advocating a return to the blindly virtuous days of yore, with its hoop skirts and unthinking conformity to social rules. Those rules were often arbitrary and harmful, and people obeyed because of the fear of disapproval. The days of ironclad doctrines and dogmatic mating rules produced just as many unhappy and constricted relationships as we see today. The difference is that in the past, divorce was virtually unheard of and generally unthinkable.

We'd Better Stay Together

In spite of all the divorces which take place today, there are even greater numbers of unhappy or dissatisfied couples who do not decide to split up their marriages. Experts maintain that only a fourth of current marriages are truly successful.[7] Many marriages are held together by an overriding concern about the extraordinary price that would have to be paid to end the relationship.

We do not mean only the financial cost, although it is substantial and a significant factor. Rather, if they separate, couples will be giving up a known situation, as bad as it may be, and entering what they fear is an even more risky and potentially dangerous phase of life. "Better the devil you know," they believe, "than the one you do not know."

Many of the marriages which remain intact are held together, then, by inertia or fears of what might happen if a change were made. "After all," they ask, "What would happen to. . ." and then they cite an unending list of worries that enter into their decision to stay married. Here are but a few of those concerns:

(1) "We wouldn't want to hurt the children. We'd better stay together for their sake."

(2) "How could we even afford to split up? It's cheaper to keep going like we are."

(3) "It's bad, but it might be a lot worse on my own."

(4) "At least I know where my socks are. I'd really hate to move."

(5) "What would our friends say? And which friends would be loyal to me? Would I still get invited out?"

(6) "My religious beliefs don't permit me to consider separation or divorce."

(7) "My career would be hurt. My image at the office would never be the same."

(8) "To tell you the truth, as bad as things are, I don't think I could do any better than what I have. It would just be the same problems with someone else."

(9) "About the worst thing I know of is loneliness. Even if we don't get along, at least I'm with somebody and have someone to talk to."

(10) "It's too scary to face the world all alone. You have to make a million decisions by yourself, and you never know what might happen. What if I got sick, or a burglar came in?"

(11) "My life style would drop like a rock. Who needs instant poverty? You can't live in half a house, and we'd both be wiped out. I've seen it happen to lots of other people. Not for me!"

(12) "The pain and anguish would be too much. It's not worth it. We have a sort of truce right now, and it should last. It has already, for 14 years."

(13) "My social status would be severely hampered. What would they say at the club?"

(14) "I can't stand the thought of the singles scene again. I'd rather put up with what I've got."

This list reflects some of the dissatisfaction endured by so many married couples. Imagine how much more unhappiness must be present for a couple to decide to actually dissolve their marriage!

The Palm Reader's Secrets

Your "heart line" runs from the outer edge of your palm across and upward to a point roughly between the index and middle fingers. If your heart line is deep and broad, it indicates a high degree of sensuality. If it curves downward instead of upward, it indicates a cooler attitude toward love and sex. A broken line means trouble in your love life.

When the hand is at rest, if your little finger is widely spread from the others, it is a clear sign that you have difficulty with intimate emotional relationships. If your little finger is thrust forward, it indicates sexual problems with your mate. A twisted pinky (little finger) indicates a liar. Beware of people who wear pinky rings— they have great problems with close relationships.[3a]

Two more palm reader's secrets: Never use your real name, and move frequently.

In sum, mismating has been the source of some of life's greatest anguish. Antique myths and modern misconceptions about mating have misled and continue to mislead millions. As society continues to change, even swinging back to a more conservative point of view, mating practices are likely to shift also.

But mating is far too important to be left to the whims and dictates of whatever mood society happens to be in at a given time!

Neither the excesses of the present nor the restrictions of the past have produced the happy, dynamic and successful relationships which lie within the potential of all of us.

It is time to look past the murky half-truths and ambiguous expectations of mating myths and biases, and move to a systematic, scientifically based strategic approach to identify, locate, attract, prosper with, and love for a lifetime the mate who is right for you.

This is the reason the Dynamic Mating System was developed.

3

The Dynamic Mating System

CINDERELLA was a loser until she got some decent shoes.

Her status as a cleaning woman, no matter how lovely and charming she was beneath the grime and soot, was probably not enough to attract the court butcher, let alone the prince.

But with her new coach and several servants, not to mention the manicure and one hellava hairdo, she gave the appearance of coming from a pretty high class background herself. In addition, she offered beauty and conversational skills, was a marvelous seamstress, and she sure could dance.

Perhaps all these qualities didn't equal what the prince had, but close enough. You know the result of her little masquerade.

The moral of the story?

Equals attract.

We call it the Parity Principle.

The Parity Principle

The Parity Principle means that *equals attract*. You have heard that opposites attract, but major research projects have shown that successful mating is usually based on similarities, not differences. If two people are approximately equal in the judged worth of their characteristics, they have parity.

Cinderella and the prince appeared to have parity and they fell in love.

She would not have had a prayer, however, if she had been ugly and her social skills had been limited to such phrases as, "Gee, what a neat house!" and "Have you been watching the Gong Show reruns?"

The next morning, when our heroine was back amidst the ashes and soot, clad again in her "just-a-working-girl" tatters and patches, she didn't even consider the possibility of going back and knocking on the castle door. What would she say? "Here I am, Prince. Great party last night! By the way, would you like to have your chimneys swept?"

She knew, without having to do much calculating, that they did not have parity. As the old saying goes, he definitely would not respect her in the morning.

The Parity Principle has been developed from the findings of a great many excellent research studies. Thousands of hours of clinical experience support its usefulness and validity. The Parity Principle deserves to be a pivotal part of the dynamic mating process.

One implication of the Parity Principle is that if you have a lot to offer, you will attract people who have a great deal to offer in return.

It also means that you can expect to get back from a relationship about the same amount you put in.[1]

Let's state it plainly:

You don't get something for nothing in mating. Relationships don't just happen through luck or fairy godmothers. You should get the mate you *deserve*. If your mate is too good for you or you are too good for your mate, the marriage will not succeed. It is as simple as that.

Parity depends on judgments, or estimates, of value. It's a lot like doing an appraisal of possessions. In the case of mating parity, however, we make judgments about the worth or value of personal characteristics. It might seem insensitive or even crass at first, but it works because it reflects reality. People make judgments all the time, but they do it badly.

Our purpose is to help you make your evaluations as accurately and realistically as possible. Your mating decisions will depend on your judgments and evaluations,

---

Fallopius to the Rescue

Men and women owe much to Gabrielle Fallopius (1523–1562), not so much because he discovered the Fallopian tubes, but because he is credited with devising the first "modern" condom. It was primitively made of linen, but was reported to be highly effective in preventing the spread of venereal disease.

A hundred years later, someone (perhaps even the mythical Dr. Condom or Contin or Condon—no one knows for sure) invented the thin sheath we know today. It was designed to prevent pregnancy, and it worked. It was made of sheep gut and tied on with a red ribbon around the scrotum. Perhaps it prevented pregnancy because people were laughing so vigorously they couldn't do anything else.

and they are too important to your life to play "Let's Pretend" and "Wouldn't It Be Nice If. . . ."

Measuring Parity

Parity depends on our own personal evaluations of ourselves and of others. We all have feelings about our own value and the comparative worth of other people. We often hear people say, "No, her background is too strong for me," or "He's too good-looking to go out with me," or "I don't think we're right for each other." These are judgments about parity or non-parity.

If two people are seen as being about equal in the worth of their characteristics, they may be said to have parity.

Most people evaluate their own mating worth and make judgments about parity on a casual and mostly unconscious level. The results can be extremely harmful. It is essential to recognize that *the enemy of effective mating is a lack of objective awareness, and knowledge is its greatest ally.*

To help you see how the Parity Principle works, let's consider the example of Al, Elaine, and Diana.

Al is twenty-seven, single, and has just arrived alone at a small dinner party. Among the guests are two unescorted women in their early twenties. Although he has never met either one before, he immediately is attracted to Elaine and is much less interested in Diana. What's happening?

It is not that Diana eats raw meat with her mouth open or forgets to comb her hair for days at a time. In fact, she is even more attractive than Elaine, and her figure is a knockout. If we were to ask Al why he was more attracted to Elaine, he likely would tell us that Diana is ". . . just great, but really, Elaine is more my type. I can tell."

Although Al probably could not tell us all the steps he went through, he has just done a preliminary screening and analysis of the two women and compared them with himself. The outcome? He and Elaine are perceived as having parity, at least at the first screening level, and he is attracted to her. Diana is seen as being "too good for me," and he passes up the opportunity to become involved.

Why didn't Al take a chance and show more interest in Diana? What did he have to lose? The answer is, for most people, a lot. Research shows that people are reluctant to show personal interest in what they see as more desirable persons. They *expect* to be rejected by them, and they fear the pain and humiliation.[2]

This doesn't mean that they wouldn't *like* to go out with someone they view as extremely desirable, however. Research has shown that when people are *guaranteed* that they will not be rejected, they select the most desirable mates available.[3] Research also shows that the most desirable date of all for most men is the woman who is considered to be "hard to get" but who, for whatever reason, is interested in him.[4]

Would Al have been rejected by the beautiful Diana? Probably, unless she perceived him as having more "mating power" than he thought he had. After all, equals attract and it works both ways.

Is it possible for the Als of the world to increase their value in order to attract the Dianas? Are there special circumstances under which Diana would be willing to become involved with Al? Is there hope that Elaine and Al will be able to rise above mediocrity in their relationship?

Yes, yes, and yes.

To understand how, we need to turn to the concept of net worth.

The Net Worth Principle

The problem for Al, Elaine, Diana and millions of others is that the evaluations and comparisons for parity are done on an unconscious catch-as-catch-can basis. Too many mistakes are made. Too many factors are not even considered. Too many wonderful opportunities for successful mating are missed. Using the Net Worth Principle, you can make your judgments and comparisons with greater accuracy and reliability.

What is net worth? We have defined it as "the sum of assets minus liabilities." When you add up the good qualities that you have, and then subtract your various liabilities, you will arrive at your net worth.

Of course, some assets are more important than others. Some liabilities will be important to you and others not. It's not a matter of just counting them.

What is needed is a system for evaluating how much each asset and liability is worth to you, personally. There is no universal scale which tells us all how much value to place on each asset or liability, because these are personal decisions, based on personal priorities.

The Dynamic Mating System is designed to allow you to find your net worth in an uncomplicated and easy-to-use process. It also reflects your own personal needs and interests.

The net worth approach represents a major breakthrough in mate selection and marriage success. The Dynamic Mating System is designed to allow you to translate your perceptions and feelings about yourself and your mate into numbers and scores for easy understanding and straightforward comparisons.

Six Mating Factors

Six essential mating factors are given indepth attention in MATING. The factors are:

- Personality
- Appearance
- Economic considerations
- Sex and affection
- Social status
- Marriage values

These factors are of critical importance to just about every mating relationship. They largely determine marital success or failure.

Under special circumstances, other variables also may be of great importance to you when you are selecting a mate or working to improve a marriage. For example, you may be especially concerned about such factors as religious attitudes and beliefs, health, or political persuasion.

These special factors can be included in your net worth calculation and considered as important for judging parity. This is discussed further in Chapter 12.

Which of the mating factors is most important? Each person will place his own level of value on each factor. For example, you may consider personality to be more important than money or social status. When you calculate your net worth, your own hierarchy of the importance of the mating factors will be reflected.

The individual's personal perceptions of the six mating factors are a very important element in the decision making process. They are at the heart of the Net Worth Principle and therefore, of course, an important part of the Dynamic Mating System.

Using the Net Worth Principle

Let's briefly rejoin Al and Elaine and what's-her-name

at the dinner party to show how the Net Worth Principle works.

Although Al would be considered reasonably handsome by most, to himself he is still the kid with the round head and jug ears he always was. It didn't help at all that Mary Alice Klott rejected him in the eleventh grade. Despite his eighteen-dollar haircut and his new penny loafers, he still gives his own appearance a low rating. He hasn't yet earned his first million, although he's amassed the sum of $6,200 and feels like he's on his way. His job title is still "minor league" as far as he's concerned. He thinks he's too shy to be a sex object, and his social status is just so-so.

In keeping with his self concept, he is attracted to Elaine and is afraid of being rejected by Diana.

What? Can he really be throwing away such an opportunity because of Diana's appearance? Isn't beauty only skin deep? Aren't the other mating factors much more important than whether someone is good-looking or

---

Bringing Home the Bacon

In Dunmow, England, there is said to be a flitch of bacon waiting for any married couple who can swear, while kneeling on two sharp-pointed stones, "that they have not quarreled or repented of their marriage within a year and a day after its celebration." The prize was first offered in 1244, but there were no takers until 1445, when the first iron-kneed couple came forward. Since then, only six more flitches of bacon have been claimed, the last in 1855.[4a]

By the way, it's not a miserly prize. A flitch is a whole side of bacon. It would not be easy to filtch a flitch!

not? Should appearance even be on the list?

Whether you agree with Al is not the issue. If attractiveness is important to him, then we must respect his concern as legitimate. Remember that assets and liabilities and the relative importance of the six mating factors depend on the individual's point of view. They are idiosyncratic.

Incidentally, Al is not alone in his emphasis on appearance. As we shall discuss later, most people put this mating factor very high on their lists of important concerns, especially when they are in the beginning stages of a relationship.[5]

But wouldn't it be worth it to Al to take a chance with Diana? We've all heard enough gloating to know that everyone loves a bargain. If Al could push himself to make a play for Diana and eventually, with a little luck, even marry her, wouldn't he be getting a very desirable beauty in exchange for what he sees in himself as having less value? The trade-off would be all to his benefit, wouldn't it?

Not really, because *equals attract*. Diana is seen as far above him, and Al doesn't even seriously consider the idea of attempting to reach that far with his limited mating power. In the same way, Al would not reach too far below his own self concept for a potential mate. The Parity Principle operates in both directions.

One of our major concerns as researchers and clinicians is the number of wonderful potential mates who are mistakenly passed up every day. Without an objective system for making judgments, far too many people marry individuals whose net worths are significantly different from their own. Without parity, their marriages are severely handicapped. The extraordinary divorce rates are not surprising in light of the fact that most marriages have been decided by luck and a lot of hoping.

Our intention is that you will be able to:

(1) Evaluate your own net worth and find an excellent mate using the Parity Principle, or
(2) Decide to improve your net worth before making any marriage choices, or
(3) If you're married, to understand the dynamics which will largely determine the success or failure of your relationship.

We will further discuss these options in later sections. In summary, the Net Worth Principle involves calculating perceived assets and liabilities and then using the Dynamic Mating System to arrive at an estimate of your net worth for mating. Your net worth will determine the kind of mate you will find most compatible.

It is important to realize that your net worth is more manageable than you probably are aware. You do not have to settle for what you have at this time. Steps for increasing your net worth, or mating power, are described in a later section.

Dangers of Net Worth Differences

"Hello. Oh, hi, sweetheart! How's your new job going? What? I *am* sitting down. What? You're going to do what? Getting married! He proposed and you . . . oh, *you* proposed . . . and he said yes. But honey, isn't this awfully sudden? Well, it seems sudden to me. Of course I'm happy! I'm delirious with joy! Well, if I don't sound exactly delirious it's because it's such a surprise. How will I tell your father and your grandmother? Listen, darling, you'd better tell me all about him. What does his father do for a living? Do they know anybody that we know?"

Parents from all walks of life worry that their marrying-age children might fall in love with someone from the "wrong" social level. They often fret that "Kristen, he's not good enough for you. He doesn't have your education or your future. He's just not our kind."

Perhaps surprisingly, parents also worry that the proposed mate might come from too high a level. They sometimes tell their kids, "Her family is too upper class for our family. They'll just look down on you, Don. You won't be happy. It can't last."

Parents may not have a systematic overview, but they have seen some of the problems that people can have if they decide, "Let's take a chance. Our love will pull us through. My parents will simply have to understand."

But to ignore inequity in relationships, as romantic as it might be, is to court disaster.

This is not to say that you cannot change your net worth. It does not mean that you have to be forever stuck at a given level. It *is* possible to increase your mating power to earn a mate with greater desirability. But you cannot expect to operate from low mating power and be able to *find* and *keep* a mate with high net worth. Both of you will soon be dissatisfied.

Conversely, you should not expect to be happy with a mate who has lower net worth, and your mate will not be happy either.

Net worths need to have parity for a mating relationship to be successful. You can work up to a higher net worth and achieve parity, or you both can start low and increase your net worths together, or you can find someone who has a similar net worth and both stay at the level you started with. In any case, equals attract and equals have the best chance of remaining mated. Unequal net worth relationships are in grave danger.

Research findings show that if two people are significantly different in net worth, both of them will be uncomfortable.[6] The one who has less value usually feels guilty or uneasy, and always has to worry that the other will discover the imbalance. This requires being on guard at all times, because the danger of rejection is real. Burdened with a sense of inferiority and inadequacy, the "lower-value" mate is likely to use some desperate ploys to maintain the marriage. Lenny Griswold is a good example.

The Saga of Lenny and Kathy

He wasn't Irish, but Lenny was lucky just the same. And everyone knew he had a double dose of good fortune when he married Kathy. She was attractive and bright, with a cheerful outlook and a pleasant middle-class background. He confided to his best friend that she was everything he could want, but in his secret moments he really couldn't understand what Kathy had seen in him.

After their marriage, Lenny became extremely controlling and set out to run Kathy's life. He discouraged all of her relationships except those he considered "safe." Her social life was decided for her, and he became more and more possessive of her time. "Concern for your safety" was how he explained what was becoming an obsession.

The Power of a Peanut

At a private girls' school in Johannesburg, students have been forbidden to eat either peanuts or peanut butter. Authorities there maintain that goobers are sexual stimulants. (Peanut butter cookies are very popular in the teachers' room.)

She was forced to account for all her time. Five minutes late, and his suspicions were aroused. When she had lunch with friends he pumped her for information. If he detected the smallest change in Kathy, he would explode at her, shouting, "You're not the girl I married!"

What were his biggest concerns? Only dimly aware of them himself, he agonized over the fear that (1) Kathy would recognize how low his net worth was and (2) she might continue to grow, leaving him even farther behind.

Imagine his alarm when Kathy said she wanted to take an evening adult education class with an older, rather mousy woman from the neighborhood. The class was "Contemporary Psychology for Women," and although he wasn't sure what it involved, he trusted the neighbor. After all, she was just as intimidated as his wife.

His threats were long and loud, as usual, but perhaps his heart wasn't in it, because she was permitted to enroll.

Six weeks later he discovered it was a consciousness raising group.

Five months later she moved out.

His lower net worth had corrupted their relationship and what had appeared to be a bargain for him turned into a nightmare of fear and suspicion. When it comes to marriage, there are no real bargains. We must deserve what we get.

George Steps Out

What happens from the opposite side of the unequal relationship? How do the mates with *greater* mating power react to non-parity marriages?

George and Louise had such a marriage. Let's find out what put their relationship into the past tense.

Louise always felt that she had been very lucky to

marry George. He was, she thought, much better looking, had so many more friends, was smarter, and he had a talent for making money that she could never hope to match.

She worried about the differences sometimes, but consoled herself with the thought, "But I love him, and he loves me, and that's what really matters." She set out to please him in all ways and rarely disagreed with him in public. In fact, she seldom voiced an opinion of any kind but contented herself with murmuring, "That's right, honey," and, "I think so too."

From George's point of view, Louise owed him a lot. He had always suspected it, but she reinforced it by continually throwing her arms around him and exclaiming, "Oh Georgie, you're too good for me!" He began to agree more and more.

Louise was, she cheerfully told everyone, just a devoted wife, happy to keep the home fires burning while she waited for her man to bring home the bacon. She read such books as *The Total Woman* and *Fascinating Womanhood* and set out to keep her guy happy by becoming a sex kitten. George was shocked to discover her at the front door slathered with whipped cream one day, and swathed in nothing but clear cellophane and a blue garter the next. She was sure that her total submission to the sexual needs of her Adonis would bind him to her with galvanizing force, or something like that. Her unconscious purpose, underneath these gimmicks, was to increase her sexual value to him.

But George attached little importance to whipped cream and garters. His marriage was, he felt, simply giving him too little, and he started to cheat on the relationship.

At first he cheated by not including her in important decisions, but she seemed contented enough making cookies in the kitchen. She didn't even appear to notice. A little later he withdrew his emotional investment in their marriage. He had, as they used to say a few years back, started "stepping out," and was very interested in a successful businesswoman who was, he thought, more his equal.

Louise was heartbroken when he told her about his decision, but she had always been very understanding. Their divorce was amicable. After all, it was she who had always said, "George, you're just too good for me."

The sad case of George and Louise illustrates a well documented fact. When one mate has greater mating power, he or she is likely to become resentful of the unfairness of the relationship. A frequent result is that the one with higher net worth will take advantage of the perceived non-parity. In fact, research shows that by far the highest incidence of extramarital affairs is found with those who feel they are "getting less than they deserve from their mates."[7]

Parity is Not Forever

Some relationships start off with parity but change when one partner increases in net worth while the other falls behind. The results are just as harmful as if they had begun with a lack of parity.

Paul and Claudia were both "winners" in high school. She had dark curly hair, Queen-of-the-Prom radiance, and he was the blonde first-string quarterback. They were a shoo-in for the yearbook's "Perfect Couple."

They were married four months after graduation, despite her parents' strenuous objections. It was obvious to everyone that they truly loved each other.

They both started college, but he didn't make the football team and he was too disappointed to spend much time with studies. After a semester he "flunked out" and decided to take a job with his cousin as a construction worker. Claudia thought about leaving school too, but her parents absolutely insisted that she continue her coursework. They were prepared, they said, to pay all her tuition and fees.

Paul's job wasn't too exciting, but he enjoyed the company softball and bowling teams, where he excelled. His friends were all rough-and-rowdy, and they put in a lot of pub time together. Claudia continued to earn good grades and increasingly thought of herself as a professional on her way to graduate school. She stayed trim and beautiful, but Paul put on 40 pounds of suet.

Within a year the construction industry hit a slump and Paul was laid off. He heard there were high-paying jobs in Alaska, and many of their arguments centered on his proposed move. By this time Paul had a girlfriend who had agreed to move with him if he got a divorce, and Claudia, who realized there was little left of their relationship anyway, had to agree. Their parity had vanished, and their relationship crumbled.

The Savings Principle

But does a couple have to have complete parity at all times for the marriage to survive? What happens when

---

Marry in Haste, Repent at Leisure

The longest case of wedding jitters was endured by Octavio Guillen and Adriana Martinez, who were married in June, 1969, when they were 82 years old. They had been engaged for 67 years.[5a]

Despite rumors to the contrary, they did not have to get married.

---

one makes a significant improvement, or if the other becomes ill or gains twenty pounds? Has parity been destroyed?

The answer is that mates can develop a reserve of good will against which they can draw. This reserve is the "savings account" of net worth theory.

The savings consist of the caring, kindnesses, help, and love which have been extended back and forth. The longer the relationship, and the more reciprocal the giving, the greater the savings will be. When the net worth of one individual falls below that of the other, parity is restored by dipping into those savings. Sometimes the savings account is inexhaustible, such as when one partner becomes terminally ill and the other continues to give and give and give without any thought of reciprocation, because it is not possible.

You may have seen examples of truly exceptional wives or husbands caring for their mates through excruciatingly difficult times. They set examples of a much deeper and more meaningful love than many couples ever achieve, and their savings accounts and "credit lines" know no limits. This kind of love is described as "dynamic love" in Chapter 4. It is based on parity and reciprocity, but with time and earned trust evolves into the kind of relationship where the partners pay less attention to who has done what for the other. They are a team, a unit, and take care of each other for the good of both.

Again, this kind of dynamic love relationship has to be earned, and it is based on parity.

All too often a marriage will not stand the stress of one partner falling behind or leaping ahead in net worth. Sometimes the one who has the higher net worth will decide to "devalue" the savings of the other. A tremendous number of divorces follow serious illnesses, a jail sentence, business losses, or other major (or even minor) setbacks. Sometimes the savings are insufficient to overcome the non-parity. Just as often the savings are discounted or devalued, and the injustice of the action is easily recognized by all of us. The next example is all too typical.

The Graduate School Syndrome

"Darling, you go to graduate school and I'll continue working. Then when you complete your studies, you can help me get my professional degree. Fair enough?"

Sounds fair enough, doesn't it? Too often, however, it's a prescription for disaster. What happens? The one who works, usually the wife, takes two jobs, works like a slave, takes care of the apartment and all living expenses, and sees the other through to the degree. With the degree in hand, however, the new graduate has a wholly new status in life and with it, more mating power. And "coincidentally" the new degree holder falls in love with someone else. It's kismet. The new love just "happens" to have higher net worth than the hard-working mate who no longer is viewed as having parity with her mate.

Fair? It's a travesty. But it also is quite predictable, and we urge couples contemplating this arrangement to preserve as much parity as possible for the sake of their relationship. They can both go to school part-time, or borrow the money, or postpone school, but whatever they do, they should recognize the importance of parity and the need to preserve it.

The Fair Trade Principle in the Mating Marketplace

But aren't there exceptions to the Parity Principle? Do people have to have identical characteristics to have a good marriage? What about the beautiful actress who marries the homely oil tycoon of decidedly advanced years? And don't we all know someone who was president of the senior class, intelligent and popular, who married a guy generally acknowledged to be less desirable?

This kind of "fair trade" is seen every day in human relationships. In the case of the actress and the tycoon, her beauty and talent are considered a fair trade for his wealth and influence. Their characteristics do not have to be identical, but their net worths must be viewed as equal or close to it.

If the actress were only beautiful, with few other assets and only enough talent to get bit parts, her mating power would be too low to interest the oil baron in a serious mating relationship. On the other hand, if he had only one small oil well, and it was now down to a wheezing trickle, she might not even introduce him to her second cousin, LaVerne.

Fair trade doesn't have to involve only economic status and appearance, however. People also do a lot of comparing of the other mating factors and make all

---

Stress in Marriage

The death of a spouse is considered to be the most traumatizing and stressful experience in a person's life. No other event in life causes so much damage to one's health and well-being. Right up there as number two on the list is divorce. Getting married ranks seventh on the list of stress-inducing experiences, high above even such shattering events as foreclosure on a mortgage, which is ranked in twenty-first place.[6a]

No, mating is never boring.

kinds of fair trades to reach parity with a potential mate. Most of the fair trading is done at an unconscious level, and if we could listen in to the hidden thoughts of people, their comparisons might sound something like this:

> "We have about the same values, except that I am much more ambitious. My parents are more normal than yours . . . in fact, your mother is a real dingbat. Their educational level is higher, and they live in a much better home. You have a much more outgoing personality. But even though I have fewer friends, I am more aggressive and can make things happen. And wait until I start earning the income I deserve! You aren't very sexy and I am a sensuality addict.
>
> I wonder . . . we're about equal now, but I think I will grow very rapidly. What will we be like together a few years down the road?"

In the mating marketplace, where deals are struck every day, you need to know what your mating power is and what kind of mate you will be able to attract. You also need to know how to make fair trades for the mutual benefit of you and your potential mate.

Just as the Net Worth Principle depends on your accurate perceptions of your assets and liabilities, so does the "fairness" of fair trade. You very likely have excellent qualities which you do not recognize as having fair trade value. As we shall see, the Dynamic Mating System includes tests and inventories for use with the Fair Trade Principle.

Factors Distorting Parity

Sometimes people end up with mating decisions they would never make under normal circumstances. But if the time is right, or the moon is full, or they are influenced by other special conditions, they can be misled into marriage to the wrong person. The young woman who provided the next example was influenced by time and special circumstances in a not unusual way:

> "I really didn't plan to get married. Jerry was crazy about me, and he kept telling me all the plans he had for us, but he really wasn't exactly my type. He was kind of wild. He'd been picked up for stealing cars a couple of times, and I know he dealt some drugs. I wanted to drop him, but what could I do for excitement then?
>
> "The thing was my parents were so old-fashioned. They wanted me in by 10:30 and I was seventeen! It just seemed like the choice was Jerry or staying at home with my mother and father.
>
> "I almost got picked up by the police a couple of times when I was with him. When you went out with Jerry you never knew what was going to happen. One night we barely escaped by climbing a freeway fence and running like crazy. It turned out the car we were in had been stolen by the kids who were driving it. It wasn't Jerry that time.
>
> "Anyway, things at home were really rotten. My parents didn't approve of him at all, and I was really upset. Even my old friends didn't want much to do with me. One night, after some screaming back and forth and my mother hit me a few times, I just couldn't take it. I said I was leaving and I went running out the door. I wasn't really going very far, but my mother yelled after me, 'You leave, and you'll have to stay out. Get back here right now or I'm throwing you out!'
>
> "I just kept going and that night I blew Jerry's mind. I said, 'Let's get married.' Just like that. We did, the next week."

As you might guess, this marriage did not last long. She had sacrificed mating parity for special circumstances and timing. She wanted to get out of her parents' home, and Jerry was available at that time. She settled for someone who was not her "type," as hundreds of thousands do every year. In a sense, Jerry received "bonus points" in his net worth for being available at the right time and under the right circumstances.

They Had to Get Married

The social pressures resulting from out-of-wedlock pregnancies can destroy any thought of parity and result in monumental mismatches. A couple with dissimilar net worths may feel compelled to marry to "give the baby a name." If special efforts are made to develop parity in the marriage, it is possible for unevenly matched couples to overcome their non-parity and build a successful relationship. It does require insight, planning, and effort, however, and these three are too often in short supply at such times. Research shows that when the bride is pregnant at the time of marriage, the relationship has less chance to survive.[8] The message is that when special circumstances (pregnancy in this case) distort the selection process (which is usually not very sophisticated anyway), the outlook for the marriage is even less favorable.

The Importance of Being Male

Why do you suppose most prospective new parents hope for a boy?[9] Because their haircuts are cheaper? They don't want to pay for a wedding? They love the color blue?

Or do they want all the advantages for their children for greater success in life, including the distinct advantag of being male?[10]

The advantages are so great, in fact, that the scientists who are working to develop a method for selecting the sex of babies are worried. They are concerned that so many people will want boys that the essential balance between the sexes will tilt toward a tremendous surplus of males![11]

Even though women have been able to vote in federal elections since 1920, they are still second class citizens in many ways. Their earnings are so markedly below their male counterparts that there is increasing concern in America about the "feminization of poverty."[12]

A recent comparison of male and female earnings for similar jobs shows some shocking discrepancies.[13] For example, women clerical workers earn 48 percent less than their male equivalents. Female sales personnel earn 33 percent less than men in comparable jobs.

Advanced education is not the great equalizer we often like to think it is. For example, women lawyers earn 29 percent less than attorneys who are men, and female physicians earn 19 percent less than male doctors. Of course, two women are now U.S. Senators, and one is a Supreme Court Justice, but it's only a beginning. As a wit once said, "We're all equal. It's just that some of us are more equal than others."

Polygamy Anyone?

About one hundred years ago, Paraguay was involved in a long, difficult war. Half of the country's population was killed. At war's end, there were only 29,000 men left for the country's 190,000 women. In order to repopulate the country quickly, the government legalized polygamy. Today, there are nearly three million people in Paraguay, and polygamy is decidedly against the law.

What are the implications of sexism for the mating marketplace? It's incontrovertible. Society gives men net worth advantages just because they are male.[14] Women have had to cope with less mating power for the lack of a Y chromosome.

Men have greater mating power for several reasons. Their parents give them more opportunities. Fathers seem especially interested in having a son who will keep the family name going. We don't throw girl babies away, or sell them cheap, as has been done throughout world history, but we as a society still look to men for leadership and power, including mating power.

As a result, women have had to work harder to compensate for their "handicap," or settle for a male with less actual net worth. It isn't fair to women or beneficial to society, but through the centuries a bonus resulting from being male has been factored into net worth calculations.

The major negotiating asset of women has been, whether we condone it or not, their appearance.[15] Research has shown that many men place an extremely high value on beauty and are willing to trade almost anything for that precious commodity.[16] In fact, a mating service recently mentioned in the Wall Street Journal caters exclusively to "Millionaires and Beautiful Women."[17]

Compared to beauty, relatively little value has been attached to the other assets that individual women have developed. For example, research shows that women who have earned Ph.D. degrees generally marry men of less net worth than do women who are beautiful. In other words, the highest academic degree that one can earn is still worth less in the mating marketplace than attractiveness. (But consider the advantage of all of the female Ph.D.'s who are also beautiful.)

No wonder parents are willing to spend so much on clothing or even cosmetic surgery for their daughters.

It's cheaper than professional or graduate school, and it will take them farther in the mating marketplace!

Sexual inequality appears to benefit men in dramatic ways. For example, a not-very-intelligent man, with average appearance and a fair-to-middling job has the opportunity to marry a much more attractive and capable female. Her actual net worth may be higher than his, but his being male gives him bonus points.

Considering all the propaganda females have been exposed to for so many years, it is not surprising that many women subscribe to the myth of greater male net worth and are willing to put up with the severe handicap placed on them. Society drums sexism into our heads from the cradle to the grave.

Single and Getting Older

Men have another mating marketplace advantage, but when you consider what it costs them, it's a dubious advantage at best! There are more women than men. It's not that men have a higher death rate, because of course it evens out in the end. But men do die earlier. Past the age of 65, for example, there are two single women for every single man.[18]

The relative scarcity of surviving men gives them even greater mating power. Many of them, relying on greater economic leverage and social status, are able to interest much younger females. The surviving older women must cope with the double disadvantage of being women and being a surplus.

Women in this age group often see their options as boiling down to only two: (1) They can attempt to attract any male, no matter how unequal his net worth, or (2) They can resign themselves to remaining single. If they are interested in marrying, however, such pessimism is unwarranted. There are, after all, a half-million single men in America past the age of sixty-five.[19] The solution rests in understanding the dynamics of the mating marketplace and using that knowledge to the full competitive advantage that insight can provide.

Gay Power

There is another important reason that fewer men than women are available for mating: more men are gay.[20] Incidentally, the potential of gay power takes on awesome proportions when you realize what happens when men, already possessing greater power in our society, team up to combine resources. They usually do not dilute their professional or financial strength by

---

Beware the Assassins

In the 17th and 18th centuries, no fashionable lady would have been seen in public without at least one gummed patch of black silk affixed to her face. Called "assassins," because they were supposed to make the heart of the beholder stand still, these patches appeared in a wild variety of sizes and shapes—stars, birds, hearts, trees, cupids, and even silhouettes of lovers.

In the French court of Louis XV, the lady's placement of the assassin indicated her intentions. If she wore it on her upper lip, it was an invitation to kiss. The center of the cheek meant she was in a gay mood, and near the eye indicated passion. Placement on the nose meant she was playful, but on the breast . . . she wasn't kidding around!

raising children, and being male, their incomes are markedly above those of females. Already their purchasing power has made them favorites of retailers, and their communal financial ability to resurrect whole sections of a city is legendary.

In sum, the advantages of being male are significant in the mating marketplace. Men are given net worth bonus points just for being male. This leads to distorted parity and creates major problems for both men and women.

The Last Person on Earth

There is one more distorter of net worth which warrants mentioning. How many times have you heard this line: "I wouldn't marry you if you were the last person on earth!"

First, let us offer our sympathy if it was really directed at you. Perhaps it was meant in jest.

Second, don't you believe it! This distorter of net worth, which we call the "perceived shortage of choices," has held many relationships together. However, it's a particularly unstable base on which to build a marriage. The reason is that when other choices become available, the ones who were "settled for" become expendable.

How to Increase Your Mating Power

By this time you know we are not touting vitamin E or powdered rhinocerous horn.

Mating power is your net worth applied to the mating marketplace. It is the sum of your mating assets minus your liabilities. It will determine the kind of mate you will be able to attract and the degree of success you have in the mating relationship.

The good news is that your mating power is not permanently fixed. It can be increased if you are willing to make an effort, or it can diminish if you prefer to be lazy and careless.

The benefits of increasing your net worth are striking, far outweighing the cost of your attempts. The reasons you might have for improving your mating power could include any or all of these:

(1) You wish to attract potential mates with the highest possible net worth. You think you deserve the best, and even if you don't qualify yet for the kind of mate you want, you intend to.

(2) You already have a mate and you would like to know how the two of you compare. You may be interested in mutual growth for a bright and successful lifetime relationship. Or you might be worried about differences between your mate and you, and want to evaluate their seriousness and learn whether they are getting larger or smaller.

(3) Many people enjoy self-improvement challenges, and you just might like to evaluate your

---

Step-families and How They Grow

It is estimated that in 1980 there were 25 million step-parents in the United States, and their numbers are increasing daily.

When parents divorce and remarry, their children gain two new step-parents and two new sets of relatives. If either remarries again, another new set of relatives is added. Some children are going to need printed programs if the trends continue.

own net worth and then improve it with the systematic and objective strategies provided as a part of the Dynamic Mating System.

(4) You might be in the process of making a mating choice among several candidates and could use the Net Worth Analysis to heip make your decision.

(5) Your choice may be between marrying now and planning to improve together, or waiting until you have increased your mating power and can attract the higher level mate you really want.

Whatever your reasons, there are two basic ways to increase your mating power. One is to make the effort necessary to make yourself more desirable as a mate. The other is to review your net worth to improve your perceptions of your assets and liabilities. Both can pay dramatic dividends.

Let's consider the self-improvement approach first and the self-concept strategy second.

The Social Stairway for Self-Improvement

The analogy of the social stairway depends on your recognition that people are different from one another, and that some people are better than others.

Now we realize that this sort of statement can be irritating or shocking, or perhaps both, because new findings, especially those from the social sciences, generally are.

At any rate, here it is again. Some people are better than others. Some are better dancers, some are better cooks, and some are better mates. More specifically, some are better for you than others.

Our point is that we do not want you trapped by the myth that everybody is equal when it comes to love. It would be romantically naive to think that love, if it is true enough and strong enough, could overcome the realities of net worth and the Parity Principle. Too many people have been too badly hurt by this tender-minded attitude.

You are aware that people are different from one another in many ways. Have you noticed how people tend to cluster at a social gathering? Those with special interests like to exchange stories and information, the best-looking ones seem to find each other, the successful people gravitate toward one another, and the shy ones hang out near each other for mutual protection.

The longer the social event lasts, the more the people with the highest social value, or net worth, will be found together. The ones with moderate net worth will talk to each other, and the wall-flowers will congregate in an inconspicuous corner.

If you imagined the social gathering taking place on a long, wide staircase, it would not take long for each step to be occupied by people with similar net worths.

Wherever you might fit on the stairway is where you have your best chance of finding the best mate for you. Remember, equals attract. You are most likely to be attracted to someone from your own level and your parity will form the basis for a successful relationship.

But let us suppose that you found yourself lower on the stairway than you want to be. You want to find a mate from somewhere "up there." Can the stairs be climbed?

The answer is yes, but you have to have the entrance fee for each level. The entrance fee is a higher net worth.

Strategies for increasing your net worth are described in later, more specialized, chapters. Suffice it to say here that the social stairway is open and wide. You might be surprised to learn that the people on the higher levels generally are delighted to welcome others. Why? So they might exchange what they have to offer for some of what you bring.

They are more than willing to trade. They are positively eager. If you have special ability, or knowledge, or talent, or beauty, or whatever your assets are for fair trade, they would like to have you share some of it with them. In exchange, they will happily share their own special assets with you. But the rule is this:

> *The more you have to bring to any relationship, the more you will be able to get in return. The less you bring, the less you will get.*

Reciprocity, the willingness to trade and share, is characteristic of successful people.

It also is characteristic of successful mating relationships.

Improving Your Self-Concept

Most people place too low a value on themselves. Perhaps familiarity breeds contempt.

At any rate, you more than likely have a higher net worth, more mating power, than you realize. If you are like most people, you have exaggerated your liabilities and given less value to your assets than they deserve. And, like most people, you probably have been overly impressed with the qualities that others have.[21,22]

An excellent way to improve your mating net worth is to take a careful look at your assets and liabilities, and discover the value that has gone unrecognized.

How do you improve your self-concept? There are several approaches available. One is to evaluate your strengths and weaknesses as objectively and accurately as possible. The Dynamic Mating System tests and inventories are developed for this purpose. They also will provide a means for calculating the net worths of others and comparing your mating power with theirs.

Another way to improve your self-concept is to spend time with people who appreciate your good points and will help you to overcome some of the bad ones. Who are these people? As much as we would like to say "family and friends," this may not be true. After all, they are the ones who have influenced (some would say *produced*) the self concept you already have.

Your social environment is a powerful influencer. Your feelings about yourself depend in largest measure on how people react to you. Your family and friends can be of value, providing you most carefully select the ones who can be helpful.

Other people who can assist you in improving your ideas about yourself include professional consultants, variously called therapists or counselors. Their purposes include helping others see themselves more clearly and making the most of their attributes. In addition, special groups are available, including consciousness-raising groups, encounter groups, therapy groups, and the like.

Increasing numbers of people who have no serious problems are using mental health consultants to learn more about themselves and their own value. As always, it is the more sophisticated in our society who make the most use of consultants.

Additionally, you should realize that your ideas about yourself will be profoundly influenced by your mate. It is up to you to find the right mate who can help you to make the most of your assets. You will, of course, reciprocate and make the effort to increase your mate's happiness and confidence.

Living with the right person can be the ultimate ego booster. If you both can work to help each other, your individual and combined net worths will skyrocket. The sky's the limit.

4

Lovetraps and Dynamic Mating

ROMANTIC LOVE is an enchanted experience. Sometimes it's a swirling blend of tempestuous passion and sublime suffering, and at other times it bubbles with a cheerful ecstasy. It comes in a wild variety of forms and shades, and it acts on people in many different ways. Despite all of its variations, however, people are quick to recognize it in themselves and in others. When you've got it, you've got it, and you know it.

But what should you do with it when you have it? Our purpose here is not to attempt the definitive explanation of what love is and is not. Rather, we are especially concerned about its influence on your mating decisions, and conversely, the influence of your mating decisions on love.

When you think about it, probably the worst time of all to choose a mate is when we are in that confused and euphoric condition called "being in love."

After all, how can we be expected to make rational decisions with our perceptions so far out of focus, feeling as if we could do anything (even deliver the moon, or

58

make pearls out of dew), admiring our true loves perched up there on the pedestals we have so thoughtfully provided?

At the same time, our glands are busy doing funny things to our bodies, rearranging our blood distribution, causing us to gasp and sigh and occasionally pant, dizzy with excitement, heart pumping and dry-mouthed—why, under these conditions it's almost impossible to decide what to have for breakfast, let alone select a mate!

But of course most of us are going to go ahead anyway and make our mating decisions while we are completely giddy with love. The Dynamic Mating System is designed to provide the objectivity and touch of reality needed at times like these, when we need all the help we can get.

Being in Love

What is being in love? Researchers have had a wonderful time working on this question. It's certainly a lot more fun than measuring the saliva output of ducks!

Their major findings are listed below. Keep in mind that not all of these changes have to be taking place for you to qualify for the "in love" condition.

If you are in love you are likely to:

- want to be with your lover as much as possible.
- be constantly thinking about your lover.
- experience intense feelings of pleasure, sometimes laughing aloud for no particular reason.
- feel extra energetic, creative, and capable of almost anything.
- discover increased and constant sensuality in the relationship.

- need more reassurance and worry more about rejection.
- be on the alert for any signs of increased or decreased attention.
- exaggerate your lover's good points and not take seriously the liabilities your lover might have.

Sounds like quite an exciting state of mind, doesn't it? Now you see why many say we are never so alive as when we are in love. Of course, they say the same thing about a close brush with death.

As a matter of fact, the love experience is quite similar, from a physiological standpoint, to being in a high risk situation, such as skydiving or riding in a motorcycle race. Many people find this physical state of excitement so exhilarating they constantly look for new partners with whom to fall in love. They are, in fact, in love with love itself.

Love Marriages

In our society, love and marriage are supposed to happen together. We are supposed to fall in love first, and then if the love is strong enough, get married. In other countries and in other times, this has not been the case at all. In fact, throughout history we find that romance was not necessary for marriage. Love and marriage weren't even considered compatible. A person was supposed to have a convenient relationship in marriage, and passion was to be found with someone else outside the home.

Come to think of it, you've seen that in your own neighborhood!

In today's America, we have an historically recent exception to the general rule. We rely on what are called "love marriages." People are supposed to marry the ones they love, and love the ones they marry.

There is no question that marriage based on love is an excellent place to *start*. However, a successful, dynamic mating relationship involves much more than just finding your true love and letting the rest come naturally.

Here are some of the problems:

(1) Most people can and will fall in love with more than one individual.[1] What should be done about the various "beloveds" who come along after a person is already married?

(2) Romantic, passionate, euphoric love does not last as long as we all might hope. What happens to our relationships when the flame flickers and goes out?

(3) Euphoric love is not known for being rational love. Many marriage mistakes are made in all the excitement.

The first problem above has contributed to much of the marital infidelity seen today. That is, when people marry the first or second person they fall in love with, what can they do about the ones they meet later? Have an affair? One reason divorce rates are so high is that many people decide to "trade in" their old models for the new mates with whom they have fallen in love.

As one client told us, "She's been a fine wife. I really hate to do this to her, but I've got to be good to myself too. She just doesn't compare to my new lady. It's no contest."

Extramarital Affairs?

We'll be blunt. We cannot join the chorus of enthusiasts who clamor for society to approve of extramarital affairs. Even though we all should recognize how powerfully exciting the "in love" experience is, a marriage relationship is a serious contractual agreement. One cannot justify bank robbery or assault on the basis of, "But it's such a thrill, Judge. I just can't help myself."

In the same way, extramarital affairs cannot be justified on the basis of the thrills involved. And we have seen too many injured people in marriage therapy to be able to go along with the contention that, "But nobody gets hurt in an extramarital affair." The victims are real. Careless attitudes are destructive. Separation and divorce are painful and costly processes for adults, their children, and society in general.

The Dynamic Mating System will help you find an excellent mate and thereby reduce the risks of dissatisfaction and boredom which cause most extramarital affairs. When you are enjoying a dynamic love relationship in a parity marriage, the complications of a "merely euphoric" love affair become much less interesting or tempting.

You will have to watch out for two timing dangers when you are making your mating choices. One is to make your decision too soon, when you are flushed with

---

When A Young Woman's Fancy Lightly Turns to Love

Among the Cashibos of Peru, the woman takes the initiative and runs the marriage show. She proposes by stealing her intended's mosquito netting and carrying it to her own bed. The couple is legally married, and remains so until such time as the wife may decide she wants a divorce. Then she throws his mosquito netting out of her hut.[7a]

love's excitement. As we have pointed out, you may find that your mate does not compare at all well with new potential mates you will meet later.

The other error is to wait too long, uncertain, holding out for perfection, but not knowing how much net worth or mating power you have, hesitating until you have let truly excellent candidates, and precious time, pass you by.

If you tend to be a perfectionist, you are liable to make the second error. If you spend too much time looking for your all-time best mate, the one who is perfect, you stand a good chance of living alone. Remember that your truly best mate probably exists somewhere, but there are many "almost best" lovers who could be fantastic for you too. In addition, the ever-changing ebb and flow of marriage means that both you and your mate will need to continually adapt and adjust to each other and new conditions. Your "perfect" mate may be much less than perfect at different stages of your relationship. Be on the alert for parity and willingness to be creative and adaptive, and you most likely will find an excellent mate for a lifetime of dynamic mating.

How Long Does Love Last?

Is it possible to stay in love for a lifetime? One answer is, "It can be, if you die from 18 months to three years after you fall in love." This is the amount of time that researchers say passionate love usually lasts.[2]

But can love last beyond the usual bounds of passionate love? Will it last past the famous "seven-year itch," let alone through the changes brought about by silver and golden anniversaries?

It is possible. And it is even possible for your love to grow deeper and more fervent as time goes by.

But to understand this, and to make this a possibility

for your life, we need to discuss the nature of love as it is usually thought of, and then introduce the concept of dynamic love.

Euphoric Love: The Thrall of It All

The public generally thinks of love as coming in two forms:

(1) the high intensity, totally absorbing euphoric love with which romantic relationships usually begin, and

(2) the mellow affection based on mature friendship which, if you are lucky, is left over when the euphoric love dies out.

Experts writing about love often divide it into essentially the same categories.[3] You should know that both have serious limitations. In fact, we refer to them as "lovetraps."

The causes of euphoric love are unclear at this point. Explanations range from instinctual/biological, which propels lovers into each other's arms for the purpose of procreation, to "critical junctures," which are points at which an individual is primed and ready for love at just the time when a lovable person is available. Critical junctures include leaving the parents' home, the death of a loved one, breakup of a love relationship, and periods of frustration and emotional turmoil.

One explanation of euphoric, passionate love considers it a form of neurosis.[4] If it is, it is certainly a popular one!

The social pressures on people to find and experience euphoric love are tremendous. We all are fed a steady diet of love themes in literature, television, and the movies, and sometimes it seems as if every song ever written is about romantic complications. Even if euphoric love were not such a thrill, the overwhelming publicity alone would make everybody want to rush out and find some.

Whatever the causes of euphoric love, it serves several functions. It is strong enough to blast contented young people from the safety and comfort of their parents' homes. It has been the catalyst for establishing new homes and having children. But because euphoric love burns out fairly quickly, it is rarely sufficient to sustain a relationship over the course of years.

The Paradoxes of Euphoric Love

Do you think Romeo and Juliet would have been such celebrated lovers if their families had not been feuding? If they had grown up next door to each other, pals from the beginning, hanging out with the same group of friends, it's quite likely that they would have married in time. But the thrill and excitement and danger (mortal, as it turned out) would have been missing, and Mr. Shakespeare's classic would be reduced to a minor subplot on *The Love Boat*.

Euphoric love is paradox upon paradox. Research suggests that it increases in intensity with setbacks and adversity. The rule of thumb is the greater the obstacles, the more the euphoric love.[5]

When parents set out to disrupt a blooming love relationship, as did Romeo and Juliet's parents, little do they realize that they are adding fuel to the intensity of euphoric love. Euphoric love thrives on challenge, difficulty, even near impossibility, and parental intervention increases the excitement of the love experience itself.

Although most young people experience euphoric love before the age of 20, it is by no means found in youth alone. It can be found in retirement communities, with joyful and giddy senior citizens fighting the obstacles and adversities in the way of their true love. What obstacles? Paradoxically, the children of older people in love often do not approve of their parents' love relationships, and inadvertantly, by objecting, strengthen and intensify their euphoric love. How the tables can turn.

What happens when the love-smitten person chases and pursues and persuades and cajoles and. . .wonder of wonders, the object of all this adoration and pursuit says YES? Does euphoric love grow deeper and stronger and more intense, as you might think? No, the intensity begins to diminish once the love object begins to reciprocate the love. What a paradox!

It's a little like the special toy that children absolutely must have, plead and beg for, pray for (usually when they are sure parents are listening), and then when the treasure is finally theirs, they suddenly lose interest. The passion to possess it has disappeared.

This is the greatest paradox of euphoric love. The emotional hurricane blows itself out, leaving the couple as husband and wife, married but without the intense emotional bond which brought them together. In a word, lovetrapped! As much as anything, they were in love with the thrill of the love experience. In all too many instances, the couple actually is left with little positive regard for each other. If they are to sustain their

---

Only in America

"All societies recognize that there are occasional violent emotional attachments between persons of the opposite sex, but our present American culture is practically the only one which has attempted to . . . make them the basis for marriage."[8a]

relationship, they must make up for lost time and quickly move beyond the waning powers of euphoric love to a more stable and enduring base. Imagine the shock and disappointment and even fear that so many people are faced with when "the honeymoon is over."

Not realizing that the nature of euphoric love is temporary and based on the excitement of challenge and adversity, they have mistaken euphoric love for personal love. When the euphoric love is gone, they mistakenly believe that their love for their mates has died, when in fact it *never was* directed at the mates as *persons*. They were lovetrapped in the thralls of the euphoric experience.

As a result of this confusion, many disillusioned and disappointed people file for divorce.

What can be done to improve the prospects that after euphoric love has run its course a more enduring love relationship will gradually replace it?

The most direct answer is to be sure you fall in love with the right person, so that euphoric love and your personal love for the individual will coincide. Unfortunately, nothing in research or experience suggests that you can turn euphoric love on or off like a faucet. There is, however, an effective and relatively painless way to select a mate with whom you can expect a lifetime of happiness *after* euphoric love is past. There also is a way for those in euphoric love to determine whether their relationship is one that can endure after euphoric love is a dim memory. We will consider these one at a time.

Choosing a Lover for Euphoric Love

The first prerequisite for selecting an excellent mate is to know what you want and what your net worth can afford. The best time to get this information is before you are in euphoric love, because, as you know, this wildly intense passion will distort your perceptions.

The Dynamic Mating System will provide you with a profile of yourself and what you desire in a mate. Carry this information with you in the back of your mind and be on the lookout for potential mates who fairly closely match your mating profile. Just as importantly, if you are serious about finding an excellent mate, position yourself so that you are likely to come in contact with potential mates who are appropriate for you.

How do you position yourself? No, we don't mean *those* kinds of positions. As we indicated earlier, it involves being where the potential mates are. An elementary school teacher who enjoys high-intensity adventures and the challenge of business is less likely to meet the right potential mate if he or she continues to teach

Lip-clapping, Anyone?

Being amorous has gone by many names throughout American history. Here are some of the terms used:[9a]

lip-clapping	lollygagging
wooing, or pitching woo	making eyes
billing and cooing	smooching
bussing	spunking up
sparking, or going sparking	mashing
parking, or going parking	necking
keeping company	petting
spooning	making out
watching submarine races	sucking face

The wild variety of terms used for intercourse is not included in this book. After all, it's dedicated to our children.

in the same school, year after year. You must be willing to be where the mates are!

By positioning yourself in the wrong environments, you can only increase the danger that you will fall in love with someone who will complicate your life and lead to unhappiness for both of you.

Let's say, however, that you have done all you can to position yourself appropriately, but . . . can you believe it! . . . you find yourself in the early stages of attraction with someone who would not have parity as a mate. What can you do?

This represents one of life's greatest opportunities to protect your future happiness. *End the relationship!*

It is a thousand times easier to discontinue unequal relationships before they develop into full-blown euphoric affairs than after the damage is done. If you have already suffered through an unsuccessful euphoric relationship, you will understand the trauma that is involved. The thrill of euphoric love is best when it has a happy ending. The key is to put distance between yourself and the one to whom you are attracted *immediately.*

It is essential that you not play games when you end the relationship. No speeches about "We can be friends anyway," or "Oh, I don't know what I'll do without you," or the other dramatic things which spring into our minds. Be simple and uncompromising, because euphoric love thrives on hope, and beating the long odds and adversity, and "We can make it, Baby." If you're in love with the love experience, you owe it to yourself to make every effort to direct all that energy output and excitement to the kind of person you will love as a person too.

If you have caught the attraction early enough, your euphoric love feelings will begin to diminish fairly shortly. It might be only days, or it could be months. In any case, you will look back and break out in a cold sweat at how close you came to making a very serious mistake.

Be sure to continue watching for the potential mates who would be best for you for a lifetime. Continue to let the environment work for you by being where the best candidates are. It will happen.

Breaking Up a Euphoric Love Affair

One of life's most difficult tasks is ending a fully developed euphoric love relationship. You have known people caught in the throes of an almost hopeless entanglement, suffering through one crisis after another, and ecstatic when it looks as if the obstacles might be overcome. One familiar theme is the plight of euphoric lovers who fight destiny for years because one is already married but cannot (or will not) give up the marriage. It is difficult to imagine the level of emotional agony that many of these people endure, literally lovesick and un-

---

Love is a Toboggan Ride

J. K. Folsom in 1934 wrote, "Love grows less exciting with time, for the same biological reasons that the second run on a fast toboggan is less exciting than the first. The diminished excitement, however, may increase the real pleasure. Extreme excitement is practically the same as fear, and is unpleasant. After the excitement has diminished below a certain point, however, pleasure will again diminish, unless new kinds of pleasure have meanwhile arisen."[10a]

Anyone like to go skiing?

able to break off the euphoric affair. It is at this serious level that a connection between euphoric love and neurosis is apparent.

Can one discontinue a full strength euphoric love relationship? Yes, but just barely. Don't underestimate how difficult it can be. You will have logged a great many mutual experiences, many of them supercharged with emotional impact. Song, food, a sunny day, just about anything can bring back visions of thrilling times together. And then, again paradoxically, when you try to end the relationship, the extra adversity it entails will increase the sense of attraction you are trying to stop. Euphoric love feeds on challenge, so the more you fight it, the stronger it is.

In this situation you are on an emotional roller coaster, going where it takes you, with never a dull moment. But of course, roller coasters don't really go anywhere, and that's the pity of euphoric love affairs with the wrong people.

You must observe one rule under these circumstances: *No long-term commitments!*

You have little control over your emotions, but you can manage your behavior. If you can accept your euphoric love feelings as not unusual and temporary, your sense of urgency will be reduced. You might as well enjoy the excitement, but keep your options open. Do not marry a person without parity. Continue to watch for potential mates with similar net worth, and remember that you can have *more* than euphoric love with the right person, as we shall see.

Euphoric Love in the Good Old Days

Earlier generations had fewer questions about what to do when euphoric love dwindled, because by the time passion was a memory, most couples were parents. The binding force of euphoric love was replaced by the glue of parental love and responsibility.

But just in case parenthood was not enough to hold couples together, society had some powerful back-up forces. Consider the plight of the woman who woke up one morning to discover that euphoric love was gone. As she sipped her morning tea she realized she was married to a man who was not only undesirable, he was downright disgusting.

What could she do? Divorced women were social outcasts. Most were unskilled and employable only at near starvation levels. Religious sanctions were rock-ribbed and implacable.

It was virtually impossible for a woman to get a divorce in most states. Even if she somehow managed to obtain one, she had little chance of marrying anyone again. Could she subject her children to the ostracism that separation and divorce would cause? What about giving up her own sex life forever?

Imagine how bad things would have to have been for a woman to attempt to get a divorce just a couple of generations ago! The "good old days" of low divorce rates and fidelity were not at all what people usually imagine them to have been. It is clear that they certainly knew how to keep marriages intact—no matter what the cost.

Today, when euphoric love burns out there is much less pressure to maintain bad relationships. Society has removed many of the sanctions and punishments of divorce, although the costs of mismating are still extremely high, as we saw in Chapter 2. The point is that euphoric love is neither good nor bad, it simply exists. It is an inadequate basis for a long-term mating relationship, but 97 percent of our population has experienced being in love by the age of twenty.[6] As we said, it's a splendid place to start, but not even a riverboat gambler should stake his future on it.

Companionate Love

An interesting book published back in 1927, entitled *The Companionate Marriage*, deals with the many joys of companionate love.[7] Since that time, others have described in a different context this more "mature" and "subdued" form of love which often follows the wild, passionate romance of new relationships. But this kind of love has major drawbacks as well. It too is a lovetrap.

Many couples have pointed to companionate love as if this relatively tepid and comfortable kind of relationship were a sort of consolation prize for having lost euphoric love.

You have undoubtedly heard people describe their companionate love in terms something like this:

> "Oh, the spark has gone out a long time ago. We like each other more now. We enjoy being around each other. It's like a couple of nicely broken-in old shoes."

Old shoes? Yes, this relationship *is* a long way from euphoric love.

In fact, it might be argued that what happens in this kind of marriage is not really romantic love at all, but more of a friendship.

Of course, friendship is a wonderful sort of relationship. One can say many fine things about friendship.

I Understand, Dear

Successful couples are twelve times more accurate in understanding what is going on in the feelings and thoughts of the partner than are the couples involved in marital conflict.[11a]

But you don't have to settle for friendship alone in your mating relationship, any more than you have to be led by your glands in euphoric love. Neither of these kinds of love is enough, as the divorce rates so clearly indicate.

Dynamic Love

Far too many people are willing to take what comes their way, willing to settle for whatever happens to them, seemingly unaware of the fantastic adventures and opportunities they are passing by. Too many people just shuffle down the pathway of life, never questioning where they are going, and following wherever it leads. If love happens to be lying across the pathway, they fall into it.

After they have fallen in love (a phrase which denotes passivity), in a while it burns out (again, they are victimized by outside forces), and they are left with the relatively weak companionate love. Companionate love by its nature also is passive—it's a way of getting by, living together, keeping company, but with no major intrinsic or extrinsic purpose.

But you do not have to settle for the limitations of euphoric love or companionate love. Dynamic love offers much more.

Dynamic love is an active process, not a condition. You do not fall into dynamic love. It is created, by both partners, never one, for it is the product of reciprocal and planned caring.

Dynamic love is characterized by a deep commitment, not only to each other, but to the relationship itself, as a valued entity, a something to be protected and treasured.

Dynamic love is based on mutual purposes and shared goals, for the individuals themselves and for the small "team" they comprise. At the very heart of dynamic love is the desire to protect the relationship and the recognition that this can be accomplished only by achieving a dynamic balance between the partners, a true parity based on mutual growth toward mutual objectives.

But, you may ask, whatever happened to "unconditional love?" Why do we have to evaluate ourselves and look for parity in net worth and build a love based on mutual objectives and an agreement to take care of and support each other?

Aren't people valuable as they are, just because they are human? Can't we expect that people will give and take love without strings attached and counting and comparing? Can't love be selfless? Has "I'll always love you, no matter what happens, now and forever" gone out of style?

As beautiful as such concepts are, and as wonderful as it would be for everyone to have, love is not cheap and easy. Like most things precious, it has to be earned.

How do you earn this precious form of love? Love comes from establishing trust, and as you know, trust comes from experience.

When two people have worked together toward mutually defined objectives, and they have given each other and their relationship the love and support and planned care they agreed upon, their trust will become total. They will no longer keep track of who has contributed what and who still owes whom. They will be a team. They will have created and earned dynamic love.

5

Seven Steps To

Dynamic Mating

THE NEXT SIX CHAPTERS contain the special tests and inventories which make up the Dynamic Mating System. In addition, they provide information about the six mating factors as they influence your mating success.

Before you grab your ball point pen and start taking the tests, however, you should remember that the Dynamic Mating System is designed as a system, a unified whole. The individual tests can be fun and useful by themselves, but they are even more beneficial as part of the overview provided by the system.

The following questions and answers are designed to give you the information you need to understand the total system and to make the best use of it.

Q. What is the Dynamic Mating System?

Of course, the whole of MATING is part and parcel of the Dynamic Mating System, but it is useful to think of it as being made up of seven basic steps. They are listed in a very condensed version below.

The Seven Steps to Dynamic Mating:

  (1) Replace mating misconceptions with effective principles based on research and clinical evidence.
  (2) Evaluate your personal net worth and the net worth of your mate (or potential mate) with the four mating principles.
  (3) Understand the three types of romantic love and what they mean for your relationship.
  (4) Decide how important each of the six mating factors is to you and whether there are other special factors which also are of major importance.
  (5) Evaluate your own marriage values and compare them with your mate's values. Determine the level of marriage readiness for you and your mate.
  (6) Calculate the Parity Score for you and your mate to determine whether your mating relationship is right for a happy, successful marriage.
  (7) Learn the basic strategies of Dynamic Mating, so that both your love and your marriage will prosper.

Q. I'm not very good at taking tests. Must I take all the tests to benefit from the chapters which follow?

Keep in mind that the tests are easy and very useful. However, there are three approaches to the testing chapters:

  1. You can postpone the tests for now and simply read the information concerning the Six Mating Factors. The tests are an added dimension which you might include at some later date.

2. You can take selected tests concerning factors of particular importance to you. The scores from tests in individual chapters can provide interesting new insights without the necessity to complete tests in all chapters.

3. We certainly encourage you to take all of the tests. In this way, you will have the advantage of composite Net Worth and Parity Scores. A major purpose of the tests, which will be lost if you don't take them, is that they will cause you to think carefully about your own mating needs, what you bring to relationships, and what you want for the future.

Q. Who can benefit from the tests?

Whether you presently have a mate or not, the tests are designed to help you enjoy the best mating relationship possible. If you are married or have a special love (or both . . . and we hope they are the same person), the test results can help you to understand and improve your relationship.

If you are single and don't really have a potential candidate for "matehood" in mind, you can take the tests appropriate for you. They will show you which kinds of mates match your needs and interests. Additionally, the system gives you important insights about where to look and how to recognize excellent candidates.

Q. What do the tests evaluate?

We will help you to evaluate yourself in six major areas which affect mating success. For those who are interested in making decisions about present or potential mates, evaluations are made in the same six areas.

The six major factors are:

1. Personality
2. Appearance
3. Economic considerations
4. Sex and affection
5. Social status
6. Marriage values.

When you complete the tests, you will be able to easily calculate your own Net Worth Score. With a little additional simple computation, you will have a Parity Score for you and any person you have evaluated with the system.

The Parity Score, with provided guidelines, will enable you to make relatively objective decisions with regard to a mate's suitability for you.

Q. What about other important assets or liabilities which aren't included in the six factors? Shouldn't they be added to your net worth?

Our six mating factors serve as a very important base for establishing net worth and parity. They are the most important variables which decide whether or not a relationship works. By using the Dynamic Mating System, however, you can expand beyond the six factors to include other variables which affect your value in the mating marketplace. This should wait, however, until you've read Chapters 6 through 11. In the final chapter, Chapter 12, you will be encouraged to consider any additional factors which influence your mating power.

Q. I am single and have no potential mate in mind at this time. After I once take the tests, can I later take them again to evaluate potential mates that I might get interested in?

Except for the dangling preposition, that is an excellent question. The answer is yes. The system is reusable.

Develop your Mating Profile now, know what qualities to look for in a mate, and when you have one or more in mind, use the Dynamic Mating System for some objective comparisons.

If you happen to be wildly in love and lack your usual good sense, it will be especially useful to you. In other words, it could find you a terrific mate or keep you out of a heap of trouble.

Q. I am trying to decide between two lovers. Can I use the system for this purpose?

Listen, Lucky, if you have several mating candidates in mind, the system will allow you to compare them and determine their relative appropriateness for you. This will involve going through the system and making judgments about each individual. With such incredibly high stakes involved, the extra effort is worth it.

Q. How can it help married people?

The system can be used to compare yourself with your husband or wife to identify various areas of difficulty and strength. Whether you have parity or not, it is important to know about it so you can plan accordingly. Later, as you progress (or have new troubles arise), you can re-evaluate your marriage and clarify what is going right and what is potentially dangerous.

In addition, the concept of dynamic mating, going far beyond the level of most marriages, should be considered as a wonderfully exciting new dimension of life.

Q. Can I really evaluate myself? Isn't it difficult to be objective?

There are many sources of insight which can be helpful in understanding yourself and determining the kind of mate who would be right for you. Think how much you could learn if you could pool all of the information about yourself that various people could provide, including former teachers, neighbors, friends, family, and lovers.

It may surprise you to learn, however, that the best source of information about you is . . . you. You know more about yourself than all the rest of the world put together. Not all that you believe about yourself is accurate, of course, but what you believe about yourself will largely determine your destiny.

Not surprisingly then, we begin by exploring with you what you believe about yourself. Will it always be objectively accurate? No. Is it always important? Yes. In fact, it is the base on which mate selection and marital happiness rests.

Q. How accurate can I expect my views to be regarding my mate? Won't my feelings for him/her distort my responses?

In the first place, we must wonder how accurate you will be if you still don't know if it's a him or a her!

Actually, the issue is not so much whether you are objectively accurate in your perceptions of your mate, but what your personal reactions are. That is, the tests help you clarify your views and organize them in an objective manner, but your personal and necessarily subjective judgments are of paramount importance.

And may we wish good luck to you and to him/her.

Q. Wouldn't it be beneficial for my mate to take the tests also, so we could compare responses?

It is a simple matter to add two additional and useful dimensions to the Dynamic Mating System:

    (1) Your mate's view of himself/herself (now you've got *us* doing it!)

    (2) Your mate's perceptions of you.

For many individuals, however, the knowledge that they would be sharing their responses with their mates would bias their answers.

The best rule to observe is this: *Before* you begin the tests, plan to keep the responses to yourself. Decide now that you are not going to share any of the individual test responses. This decision will allow you to answer test questions with greater candor and objectivity. Otherwise, you will have to take care not to injure your own pride or that of someone else you care about.

After you have completed the tests and calculated your Net Worth Score and your Parity Score, you may or may not decide to share the final results with another. This requires special consideration and thought.

The Dynamic Mating System is *not* designed to require mutual sharing of test results. Its purpose is to give you, as an individual, an opportunity to make judgments and comparisons for your best mating decisions.

Whether or not you decide to discuss the results with your mate or potential mate, your guideline must be, as always, thoughtful consideration. If you are in any way concerned that the discussion would be better postponed or forgotten entirely, you probably should not share the findings. When it comes to feelings and possible rejection of others, remember the old saying, "When in doubt, leave it out."

On the other hand, if you have a valued relationship and you think it can benefit from additional information and knowledge, it could be an extremely beneficial exercise to share results. It can provide invaluable insights and solid bases for discussion and problem-solving. If you do so, however, you should use a second book so that test results will not be contaminated. Because MATING is designed as a personal workbook and record book for future reference, your answers and computations of various scores and profiles should be kept personal and confidential.

However, you may wish to discuss the implications of test results. The sharing should be *two-way*, with your mate's Net Worth Profile and Parity Score compared and discussed along with yours. Remember that this is entirely optional and subject to your good judgment.

Q. How much time will the tests take?

The actual tests take approximately two hours, but you have no time limit. You should try to answer the test questions fairly quickly, without undue philosophical wrangling with yourself. Of course, you may become uncertain on an item or two, because they deal with some critically important issues. When you think about the six mating factors, they involve key issues for life itself, not to mention mating in particular.

The best procedure is to take your time, but keep in mind that your first reaction to a test item is probably the most accurate.

It is extremely important to work on the tests when you will not be interrupted or under time pressures. It is intended to be an exciting learning process, and it will be if you approach it correctly.

Do not attempt to do a large number of tests in one session. Space the tests over several days rather than rush through and risk missing many of the ideas which deserve your careful consideration.

Remember too that the extraordinary importance of your mating decisions for the rest of your life warrant your careful and thorough effort over a couple of days.

Q. I have problems following directions. How complicated are the instructions?

Keep in mind that the Dynamic Mating System is a

tool for helping you clarify your mating options and opportunities. Like all tools, its effectiveness depends on how you use it. Pretend that the Dynamic Mating System has a label attached to it which reads FOLLOW DIRECTIONS CAREFULLY.

The instructions are detailed, but not complicated. There are quite a few different steps to follow, but each is described carefully, with sample responses provided for clarification. It is important to follow the directions step-by-step.

As we indicated before, the extra effort to use the system carefully and thoughtfully is completely justified. Your mating decisions are among the most important you will ever make.

6

The Personality Factor

- "You'll love her! She's got a great personality."
- "You want to know why we broke up? I guess we had a personality clash."
- "We get along beautifully. He's everything I'm not. We're a perfect match."
- "We get along beautifully. He's just the same as I am. We're a perfect match."

The role of personality in mating is important, of course. We certainly could not have left it off our list of basic mating factors. But how important is it to *you*, compared with the other factors?

We have placed the Dynamic Mating System tests and inventories at the first of this and the next five chapters. After you have completed and scored the tests, the discussions will help you to understand how the various factors influence mating decisions and success. The tests will get you more involved in the issues, so that the discussions will have even greater meaning for you.

A Test-taking Reminder:

Your Net Worth Statement and Mating Scores will be based on your test responses. If you wish to evaluate only yourself at this time, take the tests designated for the *unmated*. If you want to evaluate your mate also, or potential mates, follow the instructions for *mated* respondents.

Be sure to take the tests when you are alone, and keep the individual responses confidential. Remember to let consideration and good taste guide you if you decide to discuss general results with someone later.

The M-G-B Personal Inventory was developed specifically for the Dynamic Mating System. In this test, as with the others, you are encouraged to answer as honestly as you can, because it's what YOU value that counts.

It is not necessary to worry about consistency; none of the items is repeated. If you cannot make a decision fairly quickly, pass the item and come back to it after completing the rest of the test. Work at a pace which is comfortable and pleasant for you.

Note Carefully Whether the Tests and Instructions That Follow are for Unmated or Mated.

Test 1

M-G-B PERSONAL INVENTORY
Unmated and Mated

Directions:

Step 1. Using the 1 to 5 rating scale below, you are to decide to what extent each of the following statements describes you. Place your ratings of yourself in Column A on the left (entitled "My Ratings").

Step 2. To evaluate your mate or potential mate, place your ratings of him or her in Column B on the right ("My Mate's Ratings").

Rating	How Descriptive of You (or Mate)
1	Only slightly
2	Somewhat
3	Moderately
4	Very much
5	Extremely

Test 1

M-G-B PERSONAL INVENTORY
Unmated and Mated

Column A *My* *Ratings*	Rate yourself and your mate from 1 to 5. 1=Slightly; 2=Somewhat; 3=Moderately; 4=Very Much; 5=Extremely.	*Column B* *My Mate's* *Ratings*
________	1. Able to set realistic goals.	________
________	2. Self-assured in dealings with others.	________
________	3. Seen by friends as on the way up.	________
________	4. Thought of as a well-spoken person.	________
________	5. When in need of help, can count on others.	________
________	6. Self-confident and assured.	________
________	7. An energetic problem-solver.	________
________	8. Seen by others as competent person.	________
________	9. Hard working and honest.	________
________	10. Usually calm and composed.	________
________	11. Has an open mind toward other points of view.	________
________	12. Willing to pitch in and complete the important jobs when they need to be done.	________
________	13. Able to tolerate disorder when it is necessary.	________
________	14. Busy with constructive projects.	________
________		________
Subtotal		Subtotal

_________	15. Viewed by others as very capable.	_________
_________	16. If someone else's idea is better, will change position.	_________
_________	17. Comfortable discussing feelings with others.	_________
_________	18. Enjoys discussing work and interests with others.	_________
_________	19. Doesn't worry needlessly about health.	_________
_________	20. Praises others for a job well done.	_________
_________	21. Likes to keep living area neat and uncluttered.	_________
_________	22. Can make up mind without relying on others.	_________
_________	23. Can be counted on by friends.	_________
_________	24. Good at understanding the motivation of others.	_________
_________	25. Enjoys receiving praise.	_________
_________	26. When makes a mistake, accepts blame.	_________
_________	27. Goes out of way to help people.	_________
_________	28. Enjoys things and ideas that are new and novel.	_________
_________	29. Has the stamina to see a job through to completion.	_________
_________	30. Enjoys sharing affectionate hugs and kisses.	_________
_________	31. Relentless in determination to complete those things that need to be done.	_________

_________		_________
Subtotal		Subtotal

——————— 32. Considered by friends as a   ———————
              vigorous person.
——————— 33. Friendly and outgoing.   ———————
——————— 34. Extremely loyal to friends.   ———————
——————— 35. Likes being open and honest   ———————
              with others.
——————— 36. Considered to be sharp and   ———————
              intelligent.
——————— 37. Tries to do what is right.   ———————
——————— 38. Viewed by others as hard   ———————
              working.
——————— 39. Makes good use of time.   ———————
——————— 40. Not prejudiced against those   ———————
              of other races, religions or
              national origins.
——————— 41. Sympathetic to the problems   ———————
              of others.
——————— 42. Does not give up easily on   ———————
              matters viewed as important.
——————— 43. Pleased with own intellectual   ———————
              level.
——————— 44. Creative, can find a new   ———————
              solution.
——————— 45. Enjoys showing affection.   ———————
——————— 46. Seldom monopolizes a   ———————
              conversation.
——————— 47. Optimistic about the future.   ———————
——————— 48. Good at organizing work.   ———————
——————— 49. Friendships are important   ———————
              part of life.
——————— 50. Has good insight into own   ———————
              needs.
——————— 51. Not hesitant to ask friends   ———————
              for help when needed.

———————   ———————
Subtotal   **Subtotal**

_________	52.	Apologizes when has unfairly offended someone.	_________
_________	53.	Likes diversity in life.	_________
_________	54.	Enjoys giving friends a hand.	_________
_________	55.	Is persuasive in convincing others.	_________
_________	56.	Can solve problems in creative ways.	_________
_________	57.	Comfortable with people.	_________
_________	58.	Avoids gossip which is hurtful to others.	_________
_________	59.	Can be combative at a public meeting if feels strongly about an issue.	_________
_________	60.	Has a wide range of interests.	_________
_________	61.	Tries to be a moral person.	_________
_________	62.	Reliable, keeps word.	_________
_________	63.	Can deny self those things that should be denied.	_________
_________	64.	Tries to see all sides of important issues.	_________
_________	65.	Can be counted on to help out when people need assistance.	_________
_________	66.	Can make important decisions after careful deliberation.	_________
_________	67.	Willing to accept suggestions.	_________
_________	68.	Enjoys leisure time.	_________
_________	69.	Willing and able to respect confidential information.	_________

_________   _________

Subtotal   Subtotal

_______	70.	Tries to lift spirits of others when they are down.	_______
_______	71.	Observes the law.	_______
_______	72.	Enjoys being a group leader.	_______
_______	73.	Comfortable being alone.	_______
_______	74.	Sought out and liked by others.	_______
_______	75.	Tells others what is wrong when upset.	_______
_______	76.	Is described by friends as resourceful.	_______
_______	77.	Has a well-developed conscience.	_______
_______	78.	Viewed by others as a stable person.	_______
_______	79.	Thinks things through carefully before making an important decision.	_______
_______	80.	Would fight for someone's right to disagree.	_______
_______	81.	Enjoys saying affectionate things.	_______
_______	82.	Has many hobbies.	_______
_______	83.	Is a good listener.	_______
_______	84.	Is eager to start the day.	_______
_______	85.	Described by most people as good natured.	_______
_______	86.	Has a strong determination to succeed at whatever is attempted.	_______
_______	87.	Is a happy person.	_______
_______	88.	Likes new and different situations.	_______
_______	89.	Is seen by friends as warm and affectionate.	_______
_______	90.	Has a good sense of humor.	_______

_______   _______
Subtotal   Subtotal

————	91.	Is happy to hear of the success of friends.	————
————	92.	Keeps temper under control.	————
————	93.	Believes that most people are basically good.	————
————	94.	Is patient with others.	————
————	95.	Seeks and appreciates affection from family and close friends.	————
————	96.	Finds it easy to have discussions with others.	————
————	97.	Viewed by others as mentally healthy.	————
————	98.	Has many close friends.	————
————	99.	Is trusted by others.	————
————	100.	Is involved in many activities.	————

————   ————

Subtotal   Subtotal

TOTAL SCORE   **TOTAL SCORE**
Column A   *Column B*

————   ————

(Unmated proceed to Test 2. Mated go directly to Test 3.)

Test 2

THE PERSONALITY FILTER
Unmated Only

Directions: The purpose of this test is to determine the level at which you filter out potential mates who fail to meet your minimum personality standards. Which is the *lowest* personality level acceptable to you in a mate?

1. _______ Low — The mate I select can have poor personality adjustment as long as other fair trade assets are sufficient.

2. _______ Medium — I would not select a mate, irrespective of other assets, with less than medium personality adjustment.

3. _______ High — Notwithstanding any other assets a potential mate might have, I will only accept a mate with excellent personality adjustment.

(Unmated go directly to Test 3.)

Test 3

IMPORTANCE OF PERSONALITY FACTORS
Unmated and Mated

Directions: Both Unmated and Mated answer Question A. Only *Mated* respond to Question B.

A.   How much importance do *you* attach to personality, as reflected by your motivation to improve your personal effectiveness?

(Check the appropriate level.)

Low _______ Medium _______ High _______

(Levels are described below)

B.   How much importance does *your mate* seem to attach to personality, as reflected by apparent motivation to improve personal effectiveness?

(Check the appropriate level.)

Low _______ Medium _______ High _______

(Levels are described below)

Low: Personality is not important and it's not worth making an effort to improve it.

Medium: Personality is important and it's worth making a reasonable effort to increase personal effectiveness.

High: Personality is of great importance and a major effort to achieve higher levels of personal effectiveness is clearly worthwhile.

(Unmated proceed to Scoring Instructions below. Mated turn to Scoring Instructions on page 98.)

Scoring Instructions

M-G-B PERSONAL INVENTORY
Unmated Only

Directions:

Step 1. Add the numbers in Column A in the M-G-B Personal Inventory. You will find it useful to use the subtotals for each page and then add the subtotals for the total.

Place the Total Score (page 93) from Column A here: __________ .

Step 2. Using the conversion chart below, convert your Total Score from Step 1 to Scaled Points and place a (✓) by the appropriate level:

Conversion Chart

Score from Step 1		Scaled Points	(✓)
100 to 180	=	10 Points	__________
181 to 260	=	20 Points	__________
261 to 340	=	30 Points	__________
341 to 420	=	40 Points	__________
421 to 500	=	50 Points	__________

Step 3. Look back at Test 3A (Importance of Personality Factors) on page 95.

What was your rating? __________________
(Low, Medium, or High)

Convert this rating to a number by using the following scale and *circle* the appropriate number below:

Low = 1   Medium = 2   High = 3

Step 4. Multiply your Scaled Points (from Step 2) by 1, 2, or 3, depending on which number you circled in the rating in Step 3.

For example, if you have 30 Scaled Points in Step 2 and a rating of 2 (Medium) in Step 3, then multiply $30 \times 2 = 60$.

Do your computation now and write your Personality Score here:

Personality Score ___________ .

Step 5. In the conversion chart below, you will see how your Personality Score rates in the mating marketplace. Place a (√) by the appropriate level.

(√)   *Personal Effectiveness Level*

___________   10 to 50 = Low Personal Effectiveness.

___________   60 to 100 = Medium Personal Effectiveness.

___________   110 to 150 = High Personal Effectiveness.

(Unmated now turn to the text of Chapter 6, starting on page 102.)

Scoring Instructions

M-G-B PERSONAL INVENTORY
Mated Only

Directions:

Step 1. Add the numbers in Column A (My Scores) in the M-G-B Personal Inventory. You will find it useful to use the subtotals for each page and then add the subtotals for the total. Then add the numbers in Column B (*Mate's* Scores) in the same way.

Place the Total Score
(page 93) from Column A here: _________ .

Place the Total Score
(page 93) from Column B here: _________ .

Step 2. Using the conversion chart below, convert *My* Total Score in Step 1 (Column A) to Scaled Points and note it in the space provided below. Then do the same for my *Mate's* Total Score from Column B.

Conversion Chart

Score from Step 1		Scaled Points
100 to 180	=	10 Points
181 to 260	=	20 Points
261 to 340	=	30 Points
341 to 420	=	40 Points
421 to 500	=	50 Points

Write *My* Scaled Points here: _________

Write my *Mate's* Scaled Points here: _________

Step 3. Look back to Test 3A (Importance of Personality Factors) on page 95.

What was your rating? _______________
(Low, Medium or High)

Convert this rating to a number by using the following scale, and circle the appropriate number below:

Low = 1   Medium = 2   High = 3

Step 4. Return to Test 3B on page 95.

What was your *Mate's* rating?

Mate's rating: _______________
(Low, Medium or High)

Convert your Mate's rating to a number by using the following scale; circle the appropriate number below:

Low = 1   Medium = 2   High = 3

Step 5. Next we will compute *Your* Personality Score. Multiply *My* Scaled Points (Step 2) by 1, 2, or 3, depending on which number you circled in Step 3.

For example, if you have 20 Scaled Points from Step 2 and you have circled 3 (High) in Step 3, then multiply 20 × 3 = 60S. We have placed an "S" next to the score. It stands for "Self", as this is your own Personality Score.

Do your computations and write your Personality Score here:

My Personality Score _______S (Self)

Step 6. Now we will compute your *Mate's* Personality Score. Multiply my *Mate's* Scaled Points (Step 2) by 1, 2, or 3, depending on which number you circled in Step 4.

Use exactly the same procedure you just finished in Step 5. You will notice we have placed an "M" next to the score. It stands for "Mate", as this is your *Mate's* Personality Score.

Do your computations and write your *Mate's* Personality Score here:

Mate's Personality Score ⎯⎯⎯⎯⎯M (Mate)

Step 7. The conversion chart below translates the Personality Scores for you and your mate into levels for easy comparison.

Place an "*S*" (Self) by the appropriate level for *your* score from Step 7. Place an "*M*" (Mate) by the appropriate level for your *Mate's* score from Step 8.

"S" or "M"	Personal Effectiveness Level	
⎯⎯⎯⎯	10 to 50	Low Personal Effectiveness
⎯⎯⎯⎯	60 to 100	Medium Personal Effectiveness
⎯⎯⎯⎯	110 to 150	High Personal Effectiveness

Step 8. Look at your Personality Score in Step 5 and your *Mate's* Personality Score in Step 6. Which is larger? To find your Parity Score for Personality, subtract the smaller number from the larger number and carry to your answer the "S" or "M" from the larger number.

For example, if your Personality Score is 70S and your *Mate's* Personality Score is 120M, your Parity Score for Personality is 50M. (See example below.)

Example:

$$\begin{array}{r} 120\text{M} \\ -\ 70\text{S} \\ \hline 50\text{M} \end{array}$$

Write your Parity Score for Personality on the line below:

Parity Score for Personality _________ ("S" or "M")

Step 9. The more similar your and your Mate's Personality Scores, the smaller will be your Parity Score. Perfect parity has a score of zero. The range of possible parity scores is from 0 to 140S or 0 to 140M. Remember, if the Parity Score ends with an "S", this means you have greater power with the Personality Factor than your mate. If it ends with an "M", your mate has greater power. The chart below will be helpful in interpreting your Parity Score.

Parity Score Guide

"S" (Self) Scores

Zero	Perfect parity.
10S to 40S	Excellent balance in personal effectiveness.
50S to 80S	Medium balance, but you bring greater personal effectiveness to the relationship than your mate.
90S to 140S	Dangerously imbalanced, unless your relationship includes compensating fair-trade assets. You have much greater personal effectiveness than your mate.

"M" (Mate) Scores

Zero	Perfect parity.
10M to 40M	Excellent balance in personal effectiveness.
50M to 80M	Medium balance, but your mate brings greater personal effectiveness to the relationship than you.
90M to 140M	Dangerously imbalanced, unless your relationship includes compensating fair-trade assets. Your mate has much greater personal effectiveness than you.

The Personality Factor and You

First, a reminder: Your test scores can be changed. No one wants to hear that bad news is forever. In the case of personality, dramatic progress is very possible. So, we thought it best if we started this section with some good news, particularly if your scores were low. Your scores can be changed.

No, we don't mean that you should turn back to the tests you have just taken and write in some new numbers. But you *should* think of these results, and those in the following chapters, as *guides* for you to use as you make your mating decisions.

These tests and discussions also are intended as a self review process. They show you your strengths and weaknesses, but keep in mind that your assets and liabilities are changeable. The Dynamic Mating System is designed to show you where to start.

The personality factor scores in this chapter reflect how you feel at this time about the impact of personality on your mating decisions. If you have low scores, which translate to low mating power, you do not have to settle for them. As you know, you can increase your net worth two ways:

(1) By working to improve your self-concept, and/or

(2) By increasing your assets and reducing your liabilities.

Before we discuss how to interpret your personality scores and describe some strategies for improving them, some essential background information about the nature of personality and its importance will make your scores more meaningful.

The Importance of the Personality Factor

Which is the most important factor in mating success? There is room for argument, especially if you feel it is not important for you, but most people, including many experts, point to the overwhelming influence of personality.[1]

There certainly is a great deal of everyday anecdotal evidence to support this belief. We all have observed how some couples seem to "get along" beautifully, and we often wonder which types of personalities fit together well and which combinations cause sparks to fly.

As one woman observed to us:

"It just seems like some people are made to not get along with each other. The amazing thing is that a lot of them spend their entire married lives bickering and fighting. I see couples shopping at the supermarket who probably have been married for fifty

years and they'll fight over which margarine to buy.
It makes you wonder why they got married in the
first place."

The fifty percent divorce rates of today have given us more opportunities than anyone ever wanted to look at troubled marriages close up and to hear the frequently given explanation that "We just weren't suited for each other."

Even the divorce court judges agree that incompatible personalities are sufficient cause for granting a divorce.

It is, without question, a powerfully important factor, not only in mating, but in life itself.

But what is personality? Because it is a broad concept, there are many definitions and views, and some of them are quite complicated. Several of them are pertinent to research in the mating process and should be discussed here at least briefly.

Two Schools of Thought

When it comes to personality, there is always much difference in opinion. Sometimes it becomes solidified into different schools of thought.

With regard to mating and personality, two essentially opposite positions have been taken by theorists and researchers over the years. One viewpoint is that people should look for mates whose general personalities *complement* their own, by compensating for limitations or weaknesses. The rallying cry of the people who have taken this position is "Opposites attract."

Using this approach, a shy person should mate with an outgoing person who will "make up" for his or her shyness. A highly organized person should find an easy-going and less structured mate. An emotionally volatile and fiery personality will do well with a cool, dispassionate person. According to this theory, the source of most marriage problems is that the husband and wife are just too much alike.

The rallying slogan of the second point of view is "Birds of a feather flock together."

This viewpoint holds that "Like should marry like." People get along best, according to this approach, when they are similar. Their sameness enables them to understand and identify with each other. These theorists might not say that, "The best person to marry is your own clone," but they would be tempted.

The pendulum has swung back and forth on this issue through the years. As recently as the 1950's and early 1960's there was a strong movement toward the "opposites attract" school.[3] Gradually, however, the pendulum has swung back to the "birds of a feather" camp. The evidence for either side is quite weak, but somewhat favors the similarity approach as it concerns personality.[4]

When you think about it, you can appreciate how difficult the issue has been to resolve. Two people with very similar personalities may enjoy a loving, successful relationship. But look across the street! There is a couple with distinctly different personalities, and they are equally in love and just as successful in their marriage.

And down the block and all around the nation, couples who are similar and couples who are different are found in the divorce courts too.

Let's look at the issue of marriage compatibility based on personality factors more closely, particularly with regard to what might be called personality mismatches.

Personality Mismatches

When "incompatibility" is cited as a major cause of marriage problems and divorce, what is usually meant is that two people have personality types which are in-compatible. Their interests and needs are considered so discrepant that they could not get along under the best of circumstances.

Marriage therapists often think along the lines of personality types, and they are well aware of some classic personality mismatches. Let's take the mating of a "puppy" and a "snow leopard" as an example.

The Puppy and the Snow Leopard

You've met a lot of "puppies" and you probably know a few "snow leopards" too, although as you might imagine, they are fairly scarce.

Puppies are extremely dependent on others for approval and love. Everyone loves puppies, of course, because they are so cuddly and easy to get along with. Perhaps the most important word in their vocabularies is "nice," although "neat" (as in, "Gee, that would be neat!") is a close second. They work hard to be pleasing and to do what others want them to do. In this sense, they are childlike, still far too dependent on others for a sense of being acceptable or desirable as persons.

No "Butts" About It

Which physical attribute do women most admire in men? Given a randomized list, the men guessed that women most admired "muscular chest and shoulders." Actually, that category was rated as number ten by the women. According to the London *Sunday Times*, what women admire most are buttocks.

The majority of puppies are women who have been misled by centuries of mating myths into accepting passively this immature and extremely vulnerable position in life. Despite their cheerful efforts to appear optimistic and helpful, they are most likely to feel inadequate and rather gloomy on the inside. They really are dependent. Without the approval and direction of others, preferably a strong and confident mate, they tend to flounder helplessly and miss out on many opportunities in life. Why? In the words of a puppy:

> "It wouldn't be nice to push myself forward that way. Let someone else take the leadership . . . I'm not that kind of person. I'm just little me, a good helper."

What do they wish for? A strong and decisive mate who will give them the direction and the approval they literally need. As they might say, it would be really neat.

And along stalks the snow leopard, looking powerful and aloof and fiercely independent. Just what the puppy thinks she needs for love and protection and approval.

The problem is that the proud snow leopard is not about to show tenderness and love, for this would be to give up the power barrier between himself and the rest of the world.

Most snow leopards are business executives, career military officers, professional sports figures, or in similar fields where power and competition are rewarded. Displays of love and caring are dismissed (and feared) as indicative of being "soft" and "vulnerable."

Winning is what life is all about for snow leopards.

They practice "power handshakes" and winning strategies and worry about being forceful and compelling. Cuddling is definitely not on the list of "things to do" for the snow leopard. This individual wants independence and power more than anything.

The more clinging the puppy becomes, the more distant the snow leopard becomes in turn, which causes the puppy to become even more desperate for reassurance.

Their outlook?

Surprisingly, it's quite good.

They each have strengths: puppies trust, and the snow leopards are used to succeeding. Both know how to make an effort. In marriage therapy the woman will most likely be too dependent on the therapist for a while, and the man will demonstrate vocabulary power and so forth. After a relatively short time (assuming they are reasonably well adjusted otherwise), they can be expected to make appropriate adaptations to each other's needs.

Other classic personality "mismatches" include the following:

- The careful, compulsive, steady and reliable "bulldog" and the impulsive, devil-may-care "butterfly."

- The thrills and spills powerboat jockey, grabbing for gusto and seeking high intensity excitement, married to a laid-back sailboat lover, sipping white wine and serenely gliding through life.

- The rolling stone adventurer, moving with a wanderlust from place to place, longing to see over the next horizon, mated with a tradition loving stay-at-home who values roots, the deeper the better.

It's fun to think about these different personality types, and you may know individuals and couples who fit these descriptions. It is *not* accurate to say that they are necessarily incompatible, however. Let's consider this point next.

Matchmaking Problems

The real problem with matching personality types, or deciding which personality types are incompatible, emerges when we become specific. For example, we know of many couples who have personalities which would *seem* to be mismatched, but they get along wonderfully!

Another problem is apparent when we begin to ask questions about which specific personality traits match up with others and which do not. Our clients often ask such legitimate questions as, "Since I am an outgoing, gregarious person, which personality types are best for me and which should I avoid?"

While it is possible for a consultant to help the individual client make some educated decisions about a variety of personality issues, mating choices still should not be made on the basis of specific matched personality traits. The research evidence does not justify this kind of matchmaking.

Unconscious Needs?

Another issue which often comes up when discussing personality and mating has to do with unconscious motivation and intrapsychic needs. For example, do we unconsciously seek a mate who is similar to either of our parents? Do certain characteristics, such as body type, hair color, or speech patterns, trigger our interest in a potential mate?

Researchers have examined this issue from many directions, and about the best they have been able to do so far is to draw very tentative conclusions. For example, there is a slight tendency in some people to mate with those who they think have some resemblance to their favorite parent, whether mother or father.[2]

The MATING Approach to Personality

Our own approach to personality and its influence on mating success is radically different, but justified by research findings and clinical evidence.

In our view, the issue is not so much whether specific traits of personality are similar or different. Couples can have different personality characteristics and manage superbly. They can be quite similar and manage just as well.

What matters most is the amount or level of *personal effectiveness* that each person brings to the relationship.

In MATING we are less concerned about personality as a deeply rooted psychological entity than in a more practical emphasis on day-to-day personal adjustment and problem-solving ability.

The M-G-B Personal Inventory is designed to measure your feelings about your own personal effectiveness rather than deeper levels of mental health or illness. For the sake of discussion, we will sometimes refer to personal effectiveness as adjustment or adaptation.

Personal effectiveness means being in control of your own destiny, with freedom to make choices and be your own person. In our practice of psychotherapy and through research we have discovered that effective, successful people are willing to make changes, to be adaptive. They are not passive. They are not content to accept without question whatever circumstances have dictated for their lives.

People with less personal effectiveness generally do not realize the extent to which they wait for things to happen and then react to them. If they wait for their "true loves" to come along, for example, they are more likely to be disappointed than if they make the effort to identify and locate excellent mates.

People with high scores in personal effectiveness see themselves as problem solvers, and tend to be flexible and adaptive. They are less likely to give up easily and feel overwhelmed when problems sneak up on them, as they do for all of us. As they feel more in control of their lives, they're less likely to be depressed, moody or anti-social. They tend to be more enthusiastic about their jobs, families and life in general. You can certainly appreciate why such a high premium should be placed on personal adjustment in the mating marketplace. But how do you go about assessing the emotional stability of a potential mate?

Personal Effectiveness in Mating

As important as personal effectiveness is, you might think it would be given careful attention when mating decisions are made. It usually has not been, however, for several reasons. For one thing, euphoric love has a way of distracting us from making rational evaluations of any of the mating factors, including personality.

Another important reason is that during the courtship period the prospective mates are on their best behavior. There are relatively few serious problems to contend with. Each person sees the best side of the other.

But how can you make judgments about personality when everything is going so well? It's pretty hard for a prospective mate to demonstrate low personal effectiveness when you two are enjoying a meal and looking forward to a pleasant evening together. Of course, if evidence of low personal adjustment is seen even under positive conditions, it indicates the individual is probably significantly handicapped emotionally and is a questionable candidate for marriage. . .at least to you.

It usually takes some adversity to show an individual's level of personal effectiveness. How much adversity? The greater the level of difficulty an individual can manage in a problem-solving and productive manner, the more personal effectiveness is demonstrated.

One of the best arguments for a fairly long engagement is that it offers the best opportunity to get to know each other's strengths and weaknesses, including how much personal effectiveness the other has.

Other strategies for evaluating personal adjustment include watching for examples of how the person copes with small problems, observing how his or her parents and family members manage stress, and generally being aware of the importance of this mating factor and alert for clues.

In sum, personal effectiveness refers to a person's ability to cope with stressful problems. It is a key to successful mating, and needs to be considered carefully whether you are selecting a mate or evaluating your marriage.

Personality Scores

With the personality factor as with the other mating factors, the higher your scores, the more you can and should expect from a mate. As we indicated earlier, you should get what you deserve. If you have a high personality score, you will attract others with similar personal effectiveness or those with other assets to fair trade.

If, on the other hand, your scores are relatively low in personal effectiveness, you should make every effort to increase them.

The good news is that personal effectiveness is a skill, and like all skills, is learned.

Some people are more fortunate in that they were raised in an environment which encouraged the development of problem solving ability. Many of them learned personal effectiveness quite early in life, often with deliberate instruction by their parents.

But early or late, personal adjustment is worth the effort it takes to develop.

You no doubt are familiar with the expression "Success begets success." Sometimes people say, "When you're hot, you're hot," and, "Don't stop me, I'm on a roll." What these kinds of statements mean is that winners are likely to continue winning. And it's not due to luck (at least most of the time). It comes from effort. Think again about the social stairway we discussed in Chapter 3. You climb it through increasing your net worth, but the higher you go, the more other effective people are willing to trade assets.

With regard to personal effectiveness, the more you have, the fewer problems you have (because you solve them), so the more personal effectiveness you have to devote to the remaining difficulties. It gets better and better. Success begets success.

In some respects, it hardly seems fair. Those that have get more. The people with the highest personal effectiveness attract others with equal assets and they pool their resources. Did you know that one of the most important determinants of financial success is personal effectiveness? Not only that, but it works both ways. That is, having financial success reduces or solves many problems, which translates to greater happiness and satisfaction. Research has shown that those in high positions are happier and even healthier than others. We can all stop pitying the "poor successful executive" with all his/her pressures.

As therapists ourselves, we are fully aware of the painful handicap of low personal effectiveness and its crushing effect on the lives of millions. It is a serious detriment to all facets of life, including finding a mate and building a good marriage.

We have urged you in several chapters to make the effort to increase your net worth. Let this be another reminder. To the extent that you can improve your personal effectiveness, you will solve problems more easily, attract persons with more to offer, live more happily, like yourself better . . . the advantages are unlimited. And if you marry someone with higher personal effectiveness too, and you are both interested in helping each other improve this factor, you will enjoy *compounded* dividends!

On the other hand, if your mate has personality problems, your own effectiveness is going to be severely curtailed as you are forced to cope with another source of problems rather than count on him or her to be of assistance. Let's consider the cost of low personal effectiveness in the mating marketplace next.

Personality and Fair Trade

Unlike the other mating factors, low personal effectiveness is not readily compensated for. That is, the cost of poor personality adjustment is so exorbitant that almost no one is willing to accept it in a mate. Stated yet another way, low personal adjustment is such a liability that in the mating marketplace it is almost impossible to compensate for it with other assets.

Consider the example of a young woman who comes from a wealthy family with high social status, who is attractive and sexually appealing, but who is severely disturbed. It is quite unlikely that most potential mates will be interested in her assets when she is viewed as possessing so great a liability.

On the other hand, personal effectiveness is a powerful asset. People with relatively modest incomes, appearance, and "sex appeal" can, through the force of their personalities, offset these limitations and be considered extremely desirable mates.

Parity in Personality

As with the other essential mating factors, parity in personal effectiveness is vital for successful mating. If one partner is well adjusted and the other is not, the disparity will not permit a successful relationship.

This is not to say that parity alone can guarantee a good marriage, because two troubled people, equal though they may be, are unlikely to be happy or successful whether they are together or separate. As is the case with economic status, the less personal effectiveness is brought to a relationship, the less likely it is to be successful.

Do troubled people find mates? Researchers will tell you that when they do, their mates usually have about the same level of personal adjustment.[6,7,8] Their mates are drawn from candidates who have parity. It is not that they deliberately seek out troubled partners. Rather, their net worth is too low to attract more mentally healthy ones.

This does not mean that people with low personal effectiveness are doomed to that low level forever. One of us once did some therapeutic consultation with a couple who had met as patients in a hospital psychiatric ward. They were both far from well adjusted, but their motivation to learn was outstanding. By taking personal effectiveness as a mutual goal, they helped each other to move rapidly to higher levels of personal adjustment. One discovery they made in the course of the "self-improvement project" was how little attention most people pay to the personal effectiveness factor. One implication from which they benefited greatly was that others were not really that much ahead of them! By adding even a few special skills, such as clarifying problems before working on them and asking others for assistance, they surpassed many people that they knew, including most of their relatives.

Increasing Your Personal Effectiveness

Whether you scored high, medium, or low on the M-G-B Personal Inventory, you can benefit from an increase in personal effectiveness. As therapists, we have been especially concerned about helping those with lower levels of adjustment, and have seen excellent gains come from high motivation. We also have served as consultants to people who already were quite well adjusted but who wanted to get more zest and pleasure out of life.

The point is, despite the old joke, you do not have to be crazy to see a therapist. May we suggest that you consider them as consultants and use them to increase your personal effectiveness in the same way you would use a financial consultant or legal consultant.

In addition to using therapist/consultants, a variety of books can be an excellent source of information and inspiration, as can many of the encounter and consciousness raising groups sponsored by various organizations. Another excellent resource is your mate, if you are married, or friends or trusted relatives. Keep in mind that the project has to do with improving your personal effectiveness, not necessarily working on deep-seated mental health problems.

The most important ingredient in making progress is your own willingness to try. Watch other people and observe what works. Try to be around effective people and use their techniques and insights. With effort and your recognition that the project of becoming more effective is among the most important projects anyone could undertake, you will make significant strides toward greater personal effectiveness. And you will gain not only in mating power, but in living power as well.

7

Appearance In The Mating Marketplace

MORE THAN ONE pretty face, male or female, has been traded for a life of luxury. Whether you approve of this phenomenon or not, appearance factors are significant in the mating marketplace. The issue is, however, how important they are to you.

The appearance tests, like the others, require that you respond as candidly and honestly as possible. You may evaluate yourself only, or evaluate a mate as well. In either case, follow the instructions carefully.

Again, we remind you to take the tests alone and treat the responses as personal and confidential.

Note Carefully Whether the Tests That Follow are for Unmated or Mated!

APPEARANCE INVENTORIES

Test 1

PERSONAL APPEARANCE
Unmated and Mated

Directions: This test is to be completed by everyone, both unmated and mated. Rate *yourself* on items 1 - 7 on a scale of 1 to 10, with 1 being the lowest, 5 average, and 10 highest. Item 8 is scored on a scale of 1 to 30 with 1 lowest, 15 average, and 30 highest.

1. _______ Face (Items 1 - 7 scored from 1 to 10)
2. _______ Hair
3. _______ Grooming/hygiene/complexion
4. _______ Clothing
5. _______ Body build/bearing
6. _______ Weight
7. _______ Sex appeal
8. _______ Overall physical appeal (rate self from 1 to 30)
   _______ TOTAL SCORE

(If unmated, proceed to Test 2. If you have a mate, go directly to Test 3.)

Test 2

THE APPEARANCE FILTER
Unmated Only

Directions: The purpose of this test is to determine the level at which you filter out potential mates who fail to meet your minimum appearance standards. Which is the *lowest* level of appearance acceptable to you in a mate?

1. _______ Low   The mate I select can have low appearance value if other fair-trade assets are sufficient.

2. _______ Medium   I would not select a mate, irrespective of other assets, with less than medium personal appearance.

3. _______ High   Notwithstanding any other assets a potential mate might have, I will only accept a mate with excellent physical appearance.

(Unmated go directly to Test 3.)

Test 3

IMPORTANCE OF APPEARANCE
Unmated and mated

Directions: Both Unmated and Mated answer Question A. Only Mated respond to Question B.

A. How much importance do you attach to appearance, as reflected by the motivation to improve your physical attractiveness? (Check the appropriate level.)

Low _______ Medium _______ High _______

(Levels are described below.)

B. How much importance does *your mate* seem to attach to appearance, as reflected by apparent motivation to improve physical attractiveness? (Check the appropriate level.)

Low _______ Medium _______ High _______

(Levels are described below.)

Low: Appearance is not important and it's not worth making an effort to improve it.

Medium: Appearance is important and it's worth making a reasonable effort to increase physical attractiveness.

High: Appearance is of great importance and a major effort to achieve a high level of physical attractiveness is clearly worthwhile.

(Unmated turn to Scoring Instructions page 122. Mated proceed to Test 4.)

Test 4

MATE'S APPEARANCE
Mated Only

Directions: This test examines your views of your mate's personal appearance. Respond to each item as it relates specifically to your mate. Do NOT refer to responses on earlier tests.

Items 1 through 7 are scored on a scale of 1 to 10, with 1 being lowest, 5 average, and 10 highest. Item 8 is scored on a scale of 1 to 30 with 1 lowest, 15 average, and 30 highest.

1. _______ Face (Items 1 - 7 scored from 1 to 10)
2. _______ Hair
3. _______ Grooming/hygiene/complexion
4. _______ Clothing
5. _______ Body build/bearing
6. _______ Weight
7. _______ Sex appeal
8. _______ Overall physical appeal. (Rate mate from 1 to 30.)
   _______ TOTAL SCORE

(Mated proceed to Scoring Instructions on page 125.)

Scoring Instructions

APPEARANCE INVENTORIES
Unmated Only

Directions:

Step 1. Add the numbers in the Score Column in Test 1 (Personal Appearance) on page 118.

Place the Total Score Here _________

Step 2. Using the conversion chart below, convert your Total Score from Step 1 to Scaled Points and place a (√) by the appropriate level:

Conversion Chart

Score from Step 1		*My Scaled Points*	*(√)*
8 to 26	=	10 Points	__________
27 to 44	=	20 Points	__________
45 to 62	=	30 Points	__________
63 to 80	=	40 Points	__________
81 to 100	=	50 Points	__________

Step 3. Look back at Test 3A (Importance of Appearance) on page 120.

What was your rating? _________________
(Low, Medium, or High)

Convert this rating to a number by using the following scale, and circle the appropriate number below:

Low = 1   Medium = 2   High = 3

Step 4. Multiply your Scaled Points (from Step 2 above) by 1, 2, or 3, depending on which number you circled in Step 3.

For example, if you have 30 Scaled Points in Step 2 and your level of importance score from Step 3 is Medium (2), then multiply 30 × 2 = 60.

Do your computations and then write your Appearance Score here:

Appearance Score ________ .

Step 5. In the conversion chart below, you will see how your Appearance Score rates in the mating marketplace. Place a (√) in the appropriate level.

(√)   *Appearance Level*

________   10 to 50   =   Low Appearance Power
________   60 to 100   =   Medium Appearance
                             Power
________   110 to 150 =   High Appearance Power

Step 6. Look back at Test 2 (Appearance Filter) on page 119. This test identifies the critical point at which you filter out potential mates who do not meet your minimum standards.

Which rating did you check for Test 2?

____
(Low, Medium or High)

Step 7. In this chapter only, we are going to use a special formula. Decide whether your Total Score in *Step 1* (be sure to use Step 1) is Low, Medium, or High by using the following conversion scale. Place a (√) by the appropriate level:

Total Score
(from Step 1)   *Level*   (√)

   8 to 42   =   Low   _____________

43 to 70   =   Medium   _____________

71 to 100   =   High   _____________

Step 8. Now, write your rating level from *Step* 7 (Low, Medium, or High) on the first half of Line A below. Second, write your Appearance Filter rating from Step 6 (Low, Medium, or High) on the second half of Line A.

BE SURE NOT TO REVERSE THE ORDER!

Line A. Appearance Formula: ___________________
                                                     (Step 7)   (Step 6)

Example: High   Low

(We'll use the Appearance Formula later. But now, turn to the text of Chapter 7, starting on page 131.)

Scoring Instructions

APPEARANCE INVENTORIES
Mated Only

Directions:

Step 1. Add the numbers in the Score Column in Test 1 (Personal Appearance) on page 118.

Place the Total Score here: ________

Step 2. Using the conversion chart below, convert your Total Score from Item 1 to Scaled Points and place a (√) by the appropriate level:

Conversion Chart

Score from Step 1		*My Scaled Points*	(√)
8 to 26	=	10 Points	________
27 to 44	=	20 Points	________
45 to 62	=	30 Points	________
63 to 80	=	40 Points	________
81 to 100	=	50 Points	________

Step 3. Add the numbers in the Score Column in Test 4 (Mate's Personal Appearance) on page 121.

Place the Total Score here: ________

Step 4. Using the conversion chart below, convert your *Mate's* Total Score from Step 3 to Scaled Points and place a (√) by the appropriate level:

Conversion Chart

Score from Step 3		Mate's Scaled Points	(✓)
8 to 26	=	10 Points	__________
27 to 44	=	20 Points	__________
45 to 62	=	30 Points	__________
63 to 80	=	40 Points	__________
81 to 100	=	50 Points	__________

Step 5. Look back to Test 3A (Importance of Appearance) on page 120.

What was your rating? __________
(Low, Medium, or High)

Convert this rating to a number by using the following scale, and circle the appropriate number below:

Low = 1   Medium = 2   High = 3

Step 6. Return to Test 3B on page 120. What was your *Mate's* rating?

Mate's rating: __________
(Low, Medium or High)

Convert this rating to a number by using the following scale, and circle the appropriate number below:

Low = 1   Medium = 2   High = 3

Step 7. Next we will compute Your Appearance Score. Multiply *My* Scaled Points (Step 2) by 1, 2, or 3, depending on which number you circled in Step 5.

For example, if you have 30 Scaled Points from Step 2 and you have circled 2 (Medium) in Step 5, then multiply 30 × 2 = 60S. We have placed an "S" next to the

score. It stands for "Self," as this is your own Appearance Score.

Do your computations and write your Appearance Score here:

My Appearance Score __________ S (Self)

Step 8. Now we need to compute your *Mate's* Appearance Score. Multiply your *Mate's* Scaled Points (from Step 4) by 1, 2, or 3, depending on the Importance Rating circled in Step 6.

For example, if your mate has 20 Scaled Points (in Step 4) and the Level of Importance Score (from Step 6) is 2 (Medium), then multiply $20 \times 2 = 40$M. We have included an "M" next to the score. It stands for "Mate," as this is your *Mate's* Appearance Score.

Now do your computations and write your *Mate's* Appearance Score here:

Mate's Appearance Score __________ M (Mate)

Step 9. The conversion chart below translates the Appearance Score for you and your mate into levels for easy comparison.

Place an "S" (Self) by the appropriate level for your score from Step 7. Place an "M" (Mate) by the appropriate level for your *Mate's* score from Step 8.

"S" or "M"   Appearance Level

________   10 to 50 = Low Appearance Power
________   60 to 100 = Medium Appearance
                                      Power
________   110 to 150 = High Appearance Power

Step 10. Look at your Appearance Score in Step 7
          and your *Mate's* Appearance Score in Step
          8. Which is larger? To find your Parity
          Score for Appearance, subtract the smaller
          number from the larger number and carry
          to your answer the "S" or "M" from the
          larger number.

          For example, if your Appearance Score is
          90S and your Mate's Appearance Score is
          50M, your Parity Score for Appearance is
          40S. (See example below.)

                    Example: 90S
                                  −50M
                                    40S

          Write your Parity Score for Appearance
          below:

          Parity Score for
          Appearance ________ ("S" or "M")

Step 11. The more similar your and your mate's
          Appearance Scores, the smaller will be your
          Parity Score. Perfect parity has a score of
          zero. The range of possible parity scores is
          from 0 to 140S or 0 to 140M. Remember, if
          the Parity Score ends with an "S," this means
          you have greater power with the Appear-
          ance Factor than your mate. If it ends with

an "M," your mate has greater power, The chart below will be helpful in interpreting your Parity Score.

Parity Score Guide

"S" (Self) Scores

Zero	Perfect parity.
10S to 50S	Excellent balance in appearance.
60S to 100S	Medium balance, but you bring greater appearance power to the relationship than your mate.
110S to 140S	Dangerously unbalanced, unless your relationship includes compensating fair-trade assets. You have much greater appearance power than your mate.

"M" (Mate) Scores

Zero	Perfect parity.
10M to 50M	Excellent balance in appearance.
60M to 100M	Medium balance, but your mate brings greater appearance power to the relationship than you.
110M to 140M	Dangerously unbalanced, unless your relationship includes compensating fair-trade assets. Your mate has much greater appearance power than you.

Step 12. In this chapter only, we are going to use a special formula. Decide whether your Total Score in *Step 1* (be sure to use Step 1) is Low, Medium, or High by using the following scale. Place a (√) by the appropriate level:

Total Score
(*Step 1*)		*Level*	(√)
8 to 44	=	Low	__________
45 to 80	=	Medium	__________
81 to 100	=	High	__________

Step 13. Now decide whether your *Mate's* Total Score in *Step 3* is Low, Medium, or High by using the following scale. Place a (√) by the appropriate level:

Total Score
(*Step 3*)		*Level*	(√)
8 to 44	=	Low	__________
45 to 80	=	Medium	__________
81 to 100	=	High	__________

Step 14. Write your rating level from Step 12 in the first space on Line A below. Second, write the Importance of Appearance rating from Step 5 in the second space on Line A. Third, write your *Mate's* rating level from Step 13 in the third space on Line A. BE SURE NOT TO MIX UP THE ORDER.

Line A. *My* Appearance Formula:

	(Step 12)	(Step 5)	(Step 13)
Example:	Medium	Low	High

(We'll use the Appearance Formula later. But now, turn to the text of Chapter 7 on the next page.)

The Importance of Appearance

If you rated appearance as of major importance to you, you are definitely in the majority. Research shows that nearly everyone recognizes and places a high value on attractiveness.[1,2] These are opinions, of course, with no "right" or "wrong" ratings.

But whether you rated appearance as high, medium, or low, you should be aware of the extraordinary impact that appearance has on our lives. It is not only a key motivator in mating decisions; appearance plays a lasting and very influential role in just about every facet of a person's life.

Attractive people are given every advantage. Unattractive people have to struggle against a significant handicap which is real and extremely limiting. Skin deep though it may be, beauty is seen as extremely important and appearance is made a dominating factor in the lives of nearly everyone.

Beauty is considered such a valuable characteristic, in fact, that many people, especially men, will fair trade almost any net worth asset they have to acquire it in the mating marketplace.[3]

"You can't judge a book by its cover" may be true, but it doesn't keep people from trying. In fact, in the same way that a book's cover can help make it a best seller, or consign it to oblivion, so does the physical attractiveness of a person influence the amount of success or failure which will be experienced.

People are judged by appearance from birth until they die. "What a beautiful baby!" we exclaim, and we pick up and hold attractive babies more than we do the unattractive ones. "Your child looks like a model," people coo, and they make plans to include the youngster in the next birthday party. All along the line, attractive people are blessed with extra benefits. When they grow old they hear, "You look wonderful."

In fact the judgments continue at the funeral! "Doesn't he look splendid lying there?" "She is as beautiful as she was twenty years ago!"

The Power of Beauty

It isn't so much that the evaluating of appearance is wrong. People cannot be expected to overlook differences or not to make comparisons. The problem is that appearance has been given inordinate importance, often overpowering other mating qualities.

Even at the nursery school level, children who are considered physically attractive are liked better than the youngsters considered unattractive. Unattractive boys are thought of as more aggressive and frightening.[4]

Being teased about appearance in childhood has an indelible effect, as you are likely aware.[5] It is especially painful and long-lasting when a trusted and respected adult, such as an uncle, teacher, or parent blurts out something like, "Oh, here comes that funny looking little guy we were talking about." Other children can be cruel, but careless adults are even more destructive.

The effects of early teasing about appearance have handicapped people for their lifetimes and brought millions to psychotherapy.

What Attractiveness Can Do For You

The positive advantages of attractiveness are rather astounding. People are chosen leaders or given promotions because they "looked the part." President Harding was touted as the candidate who ". . .looked good, like a President should." It is not a coincidence that presidential candidates all seem to dress alike, and when they need to be their most persuasive, they put on their "sincere-but-powerful uniforms," consisting of a blue suit, white shirt, and maroon tie.

More than one aspiring corporate officer or banker has deliberately turned his (or her) temples silver-gray or dressed in somber pinstripes to achieve the "look" of a successful executive. And the knowledgeable ones always work to be as physically attractive as possible, with their weight and clothing and posture and hair styles and all the other appearance factors as much under control as they can manage.

It is not an accident that advertisements for clothing refer to the "statement" behind the fashion. How we dress does tell the world something about us. So do posture and gestures. In the same way that smart people learn how to speak well, they learn to communicate and influence in nonverbal ways too. Making an effort to be attractive tells people a great deal.

Is attractiveness that important, even in the upper echelons of power? Indeed it is, because most successful people know about the power of attractiveness, and they trade on it. They are generally aware, for example, that people who are considered attractive are *thought* to have a great many other desirable traits that they may not actually have.[6]

What are the presumed good qualities? The list is long and impressive. Various studies have found that attractive people are considered to be:

- smarter,
- more popular,
- better educated,
- more interesting,
- in more prestigious occupations than unattractive people!

Not only that, but the various studies also showed that people with highly regarded appearance *also* were thought to be:

- more ambitious,
- stronger,
- more sexual,
- warmer and friendlier,
- in a higher socio-economic class, and
- more likely to succeed!

Quite a list. Are you convinced of the power of attractiveness? Those same studies *also* found that attractive people were judged to be:

- happier,
- more responsive,
- more poised,
- more modest,
- more sociable,
- more kind,
- and even more moral!

Small wonder that knowledgeable people do everything in their power to be as attractive as possible. The dividends are astounding!

Probably because they have been given extra attention and advantages since they were young, people who think of *themselves* as attractive see themselves in more positive terms.[7] They consider *themselves* to be more:

- likeable,
- assertive,
- successful,
- conscientious,
- popular,
- and even more intelligent than
  the average person.

Appearance in the Mating Marketplace

How does appearance influence the mating marketplace? It is especially important in the beginning stages of relationships, when people notice and become attracted to each other. It is by far the most important mating factor when a relationship begins.[8]

If you are a physically attractive person, you will find yourself sought after as if you were a treasure. Attractive people have the desirable qualities which allow them to compete successfully in what we refer to as "open settings." Less attractive people are more likely to find desirable mates in "closed settings."

Open and Closed Settings

When you first meet someone in a social situation which has no specific rules or expectations, you are in an open setting. You have many options. You are free to get to know each other or not, and you are not required to follow any special rules. You can even leave if you wish.

The great advantage of the open setting is its sheer numbers of people. You will find more competition in the open setting, however, and less opportunity to get to know people well. Examples of open settings include attending a political rally, shopping at the mall, going to a dance or singles bar, and lying in the sun at the beach.

How do people make judgments about each other in the open settings? They use the information they have, which usually isn't much. They can't, for example, really know much about the values or personalities of the potential mates they see around them. About the only attribute they have to judge is attractiveness. In this sense, then, attractiveness serves as the first "filter" through which potential mates must pass.

Obviously, if there is not enough attractiveness to get a person past the first filter, to even have a *chance* to talk and compare interests and discover some other wonderful qualities, then the open setting should be avoided. Oh, we don't mean people with less attractiveness shouldn't go shopping or to the beach. It's not your best place to find a mate, however.

The closed setting is more promising for the less attractive, because it provides ongoing contact over a greater period of time. Expected behaviors are prescribed by the activity, and the interaction gives each person an opportunity to get to know the other better. Closed settings include working in an office, belonging to a special interest group, living in an apartment complex, contributing time to a volunteer organization, enrolling in a small class, or joining a church.

People with low attractiveness scores should emphasize closed settings for finding a mate, so that they have a chance to demonstrate some of their other assets before they are eliminated from contention by the appearance filter.

Over the last forty years or so, one consistent conclusion has come from research on affection: We become very fond of those people we come to know well and with whom we spend a great deal of time. If they are potential mates, we are very likely to become romantically involved.[9] This means you will have a tremendous advantage if you can position yourself in a closed setting so that you see potential mates regularly. It has major implications for how to spend your time and maybe even for where you live and work.

Linda Breaks the Routine

Consider the case of Linda. She is a well-groomed single woman of 25 who does not consider herself to be particularly attractive but who would love to meet the right person and get married. Her problem is that she is rarely around eligible males. As an elementary school teacher she works with 19 women, six of whom are also single, and only five men, all of whom are either married or considerably older than she is. Her job would provide an excellent closed setting for *males* who are looking for a mate, but holds little promise for her.

In the last year, Linda has been spending one evening a week at singles bars. Initially she was reluctant to go, but she convinced herself that she had to break with her regular routine or her love life would never improve. Unfortunately, things rarely work out for Linda at singles bars. The few men who interest her aren't interested in what she considers to be a "worthwhile relationship."

Put Your Money Where Your Face Is

How people spend their money tells much about what they value. We Americans spend staggering sums on "looking good." The following list gives some idea of the amounts spent for appearance:

- Over 3½ billion dollars per year for cosmetics (lipsticks, face creams, makeup, etc.)
- Over 2 billion dollars for hair products (shampoos, hair sprays, rinses, etc.)
- Over 1½ billion dollars for shaving products (blades, lotions, preparations, etc.)
- Over ¾ billion dollars for fragrance preparations (perfumes, toilet water, cologne)
- ⅓ of a billion for hand products (nail polish, lotions)[12a]

These alone total more than the expenditures for *everything* in such countries as Argentina, Belgium, and Norway.

Every week Linda tells herself that the bars are a lost cause for her and that she shouldn't bother to go back. But hope springs eternal and by the next week she and her girl friends are usually ready to give it another try.

We cannot be at all optimistic for Linda until she develops a new approach to finding a mate. Her present strategy gives short shrift (is there such a thing as long shrift?) to her assets and maximizes her liabilities (low appearance value). The challenge for Linda will be to find a closed setting in which eligible males are available who will appreciate her assets. After all, her social status as a teacher and her economic security are major pluses in the mating marketplace. Where to begin?

Linda should be aware of the need to position herself to greater advantage. Are her living arrangements conducive to meeting eligible mates? Is she living at home or in some other setting in which she's unlikely to meet many men? Which organizations can she join to which eligible males belong?

The key is not to seek out males with a scatter-gun approach in open settings such as singles bars. This will only dilute the time available without giving her the advantages of the closed setting. Far better for Linda that her strategy be to concentrate on one or two more daring options.

She may elect to take a one year leave of absence, if she can afford it, and return to graduate school. There she could enroll in or just audit courses in such largely male preserves as graduate business or engineering. It would be best, of course, if she had an interest in the courses themselves. Or she may decide to make a major push toward elected office with a local organization in which she is interested.

Another excellent option is to become involved in a business which has potential mates in abundance. One enterprising young woman of our acquaintance began her own singles club! What Linda should recognize is that her chances are reduced the longer she remains in an environment which does not have the potential to provide her with parity mates.

Improving Your Appearance Value

What should you do if your appearance is a liability in the mating marketplace? The answer is obvious, but just as obviously, a great many people have not acted on it. Less attractive people should make every effort to improve their attractiveness.

There are two arguments against this, but neither is very convincing. One is that people should not have to change what nature or God gave them. You might hear, "This is the way I am, and I will stay as I am. People will have to love me for me."

Brave words, and we wish them well. It's the same logic used, however, when a person refuses to modify any liability. The consequences of unattractiveness are predictable and potent, and no one should have to suffer them needlessly. An effort to improve appearance is definitely worthwhile.

The other argument is that unattractive people are that way by birth or accident, and there is nothing which can be done. This point of view is unfortunate, because it confuses attractiveness with natural beauty, including

---

Break the Mirrors!

A survey by *Psychology Today* showed that nearly one half of the respondents were dissatisfied with their teeth, one in four didn't like their complexions, and one fifth were unhappy about their noses.

symmetry of features, a straight nose, and the like.

Actually, attractiveness has as much to do with grooming, posture, voice, cleanliness, manner of dress, openness to communication, weight, friendly self-confidence and the way one moves and talks as it does with attractive features. This is not to say that excellent grooming will equal the assets of a gorgeous model or handsome film star, but it makes the difference a lot smaller.

Proper diet, attractive and flattering clothes, attention to details in grooming . . . suggestions for all of these are provided in scores of articles and books, and they are worth taking seriously. Attractiveness at least *starts* with being skin deep.

Another useful strategy is to enlist the help of people you trust who are attractive themselves. The more a person associates with physically attractive people, the greater gains he makes in prestige.[10] It pays to have attractive friends. Keep working on improving your net worth and your specialty, as we discussed in Chapter 3, so that you will have desirable assets to trade.

Finally, if you have rated yourself low on the attractiveness scale, you can expect that potential mates with similar levels of attractiveness will be interested in you. Research shows that men and women are reluctant to approach others who are significantly different from themselves in appearance. As approximately nine percent of the adults in our society consider themselves to be low in attractiveness, there are many others with whom you will have parity in appearance.[11] The fact is that people generally choose mates who are about as attractive as they are.[12]

One word of reminder is appropriate here, whether you have rated yourself as low, medium, or high in appearance. Some excellent qualities are often hidden from the public because too few ever looked past the exterior of less attractive individuals. It is well worth your while to review the assets of potential mates who do not necessarily possess natural beauty. There are many kinds of "isms" in our society which are limiting and destructive. They include racism, sexism, and ageism. We are convinced that *appearancism* should be added to the list.

Appearancism is the arbitrary bias or discrimination against physically unattractive people based on appearance factors over which they can exercise little control.

Appearancism is widespread in hiring and promotion practices. It unfairly penalizes people for their lack of natural attractiveness. It is unjust and morally repugnant. Why should cheekbones and hairlines be so significant, when they have nothing at all to do with a person's ability or effort or character!

Even though appearancism is unfair, it is justifiable to evaluate and compare people based on appearance factors which can be managed through reasonable effort.

The point is that much of appearance is determined by variables the individual can control. Attractiveness is, as much as anything else, looking after one's self. When a person does not make the effort to be well groomed, it shows carelessness or the inability to control even this small part of one's own destiny.

If you have rated yourself low or even medium in appearance, you might want to consider again whether you have taken advantage of the wide variety of appearance-improving processes which are so readily available. It is not necessary to go through life handicapped by unattractiveness which can be changed. You are likely to find that the returns from your efforts to increase your attractiveness will be both abundant and rapid.

Interpreting Your Appearance Scores

This chapter uses a special method for helping you analyze your test results. Individual profiles have been developed which correspond to your Appearance Formula at the front of the chapter. Follow the instructions carefully. Note whether the instructions are for Unmated or Mated.

Unmated:

1. Return to page 124, Step 8. Note your Appearance Formula on Line A and place it here:

    _______ _______
    (Step 7) (Step 6)

2. Use only the first letter of the two words (L for Low, M for Medium, and H for High) in your Appearance Formula. Locate your formula below and turn to the page indicated.

    LL Page 143   LM Page 144   LH Page 144
    ML Page 145   MM Page 146   MH Page 147
    HL Page 148   HM Page 148   HH Page 149

Mated:

1. Return to page 129, Step 12. Note your Appearance Formula on Line A and place it below:

    My Appearance Formula

    ____
    (Step 11)   (Step 5)   (Step 12)

2. Use only the first letter of each of the three words (L for Low, M for Medium, and H for High) in the Appearance Formula. Locate your formula below and turn to the page indicated.

    LLL Page 150   LLM Page 151   LLH Page 151
    LML Page 152   LMM Page 153   LMH Page 154
    LHL Page 154   LHM Page 155   LHH Page 155
    MML Page 156   MMM Page 157   MMH Page 157

UNMATED APPEARANCE PROFILES

LL Profile

You have rated yourself as low in physical appearance and as willing to accept a mate with low appearance value. Most people with this LL combination have been discouraged and have given up, saying in effect, "I just don't care about appearance." In a few cases this rejection of the importance of appearance is related to a philosophical or religious reaction to the excesses of "worldliness."

You should recognize the extreme importance that most people place on appearance and review whether your own rejection of its importance is worth the price. Keep in mind that appearance almost always can be improved dramatically with effort.

Closed settings should be emphasized and selected with care. You need an opportunity to show your other assets to potential mates, and the best place to do that is in a work situation or other setting where you are with desired possible mates on a day-to-day basis.

Remember the importance of developing a specialty or two which will offset appearance liabilities. You should find a review of the sections dealing with the social stairway and improving self-concept very useful. You might also ask yourself whether your mate's appearance really means so little or are you so concerned about the possibility of being rejected that you avoid even considering a relationship with someone who is significantly more attractive. Turn to page 167.

LM Profile

Your desire for a mate with medium appearance can be a positive sign of optimism and a healthy self-concept. Very often people who perceive themselves as having low attractiveness are so fearful of rejection that they seek parity by approaching only those potential mates who also are less attractive than average. This can be a disservice to both parties; no one should be "settled for," but chosen for positive reasons.

As long as you view your own appearance as poor, you are likely to have a great deal of difficulty interesting more physically attractive people. Unless you have other major assets to trade in the mating marketplace, you will be at a serious disadvantage. The remaining tests will allow you to calculate your other net worth assets and decide which should be emphasized.

Your most immediate task should be to re-evaluate your own appearance score and determine which characteristics can be most readily improved. In addition, reread the section in this chapter on open and closed settings, and in Chapter Three look again at the section on improving your mating power. Emphasize a closed setting rather than an open one. And keep your healthy optimism. Turn to page 167.

LH Profile

Your appearance profile suggests a major mating challenge ahead for you. Because appearance is such a major influencer of mating choices, you may find that the attractive mates you want are not going to be initially attracted to you. Your task is to get past the first filter or screening stage, so that your other assets can be appreciated.

The following recommendations should be considered:

(1) Rethink your own appearance value. Have you accurately rated your attractiveness as you usually feel about yourself, or were you especially discouraged when you took the test?

(2) Which appearance factors lend themselves to improvement? With the high aspirations you have to compete for the most attractive, you should make the most of your own appearance.

(3) Emphasize the closed setting opportunities rather than open settings, so your other assets can be more readily recognized.

(4) Give a great deal of attention to increasing your mating power through the development of nonappearance assets. As you continue through MATING, be alert to those areas in which you can develop high levels of recognition. Turn to page 167.

ML Profile

This is a relatively rare profile. Most people who place a low value on a mate's appearance also see themselves as lacking attractiveness. When respondents see themselves as of average appearance, as you do, they usually are interested in mates of at least average appearance also.

Depending on your reasons for the ML rating, you can have a distinct advantage in mating or you need some extra review and effort to offset potential problems.

If you are of medium attractiveness and are willing to accept a mate with low appearance, you will enjoy the widest range of potential mates. Many will have been overlooked by others because of their limited attractiveness and will have exceptional assets to offer. Many of them will value your attractiveness and be willing to fair trade other highly valued assets to have you as a mate.

On the other hand, your lack of interest in a mate's appearance may reflect a poor self-concept and fear of rejection. If this is the case, mating problems are likely to emerge, as low self-concept usually leads to non-parity relationships.

Either an open or closed setting is appropriate. Do not forget the benefits of developing other assets. Turn to page 167.

MM Profile

The MM Profile has much to recommend it. Your self rating and willingness to accept a mate of medium appearance bodes well for a parity relationship. Your view of yourself suggests an adequate self-concept, and your interest in a mate of at least medium appearance suggests no unreasonable fears of rejection.

However, this profile is fairly neutral. In a highly competitive open setting you may be at some disadvantage. You would be well advised to review the items in Appearance Test I to consider whether you can increase your appearance assets. Keep in mind how much power is attached to attractiveness in the mating marketplace. Do not overlook the importance of any special talents or skills or knowledge which can be developed. Additional mating factors and their value for you are described in the chapters which follow. Turn to page 167.

MH Profile

You see yourself as being of medium attractiveness but are willing to accept only highly attractive mates. This profile is both fairly common and healthy. It usually is based on a realistic assessment of personal appearance and a self-concept which is sufficiently strong to look for greater attractiveness in a mate without great fear of rejection.

If you are male, you have the extra advantage of the sexual bias which makes men more likely than women to have mates who are more attractive than they are.

For you to realize your goal of a very attractive mate, you need to be very much aware of the Fair Trade and Net Worth Principles. From just the standpoint of appearance, you lack the assets to attract a beautiful or handsome mate. The following steps are in order:

(1) Review the items on Appearance Test 1 to determine in which appearance areas you can most readily improve.

(2) Concentrate on increasing your mating power through the development of non-appearance assets which can be used to achieve parity with an attractive mate. The chapters which follow will be helpful in identifying those asset factors.

(3) A closed setting will be more beneficial, but it must be a situation which includes attractive potential mates. From this standpoint, fashion photographers and movie producers have it all figured out.

Keep in mind that this is an essentially positive profile which gives you an excellent start in the Dynamic Mating process. Turn to page 167.

HL Profile

The HL Profile is quite rare and usually found with people who have a highly developed sense of priorities which makes attractiveness not very important to them. Sometimes people with this profile, however, have not really thought through the implications of attractiveness.

If your self-perception is accurate, you have an outstanding opportunity to mate well. You can select from the widest range of potential mates, because your own attractiveness will give you sufficient mating power to pass the initial filter and be given serious attention. If you wish, you can demand very major assets in other areas in exchange for your superior attractiveness.

Almost all people with high attractiveness are interested in others with similar appearance levels. It might be dangerous for you to mate significantly below your own appearance status, inasmuch as lack of parity may become destructive later on. Your mate must bring other highly valued and durable assets to your mating relationship.

You can be expected to compete very successfully in an open setting, which gives the advantage of exposure to many more potential mates than does a closed setting. Do not overlook the benefits of building your net worth with other specialties. You may not need them for mating, but they will add to your general ability to influence your own destiny. Turn to page 167.

HM Profile

There are excellent advantages for the fortunate few who are highly attractive but willing to accept a mate with lower appearance value. As you are aware, appearance is among the most powerful assets available in the mating marketplace. It is so valued that you will find many excellent candidates for mating who will be extremely interested in you. Your willingness to accept a mate of medium attractiveness means that you can exchange or fair trade for other assets you may value much more highly. Your mate is likely to place a very high value on your appearance and will be willing to trade some truly significant assets to have you. This in itself is not a problem, but you also must be valued for your other assets or the relationship will be in jeopardy with the passage of time. You cannot afford to be exploited for your attractiveness. Turn to page 167.

HH Profile

Most people with highly rated attractiveness also place a premium on similar attractiveness in their mates. Equals attract. The Parity Principle prevails. What does this imply for you in the mating marketplace?

Because of your high expectations for your mate, you will limit the number who will be appropriate. In addition, people with great attractiveness often trade their high appearance value, which is a powerful asset, for the greater economic or social assets of less attractive mates. In other words, your attractiveness alone may not be sufficient to interest other very attractive potential mates who may be more interested in other assets such as high economic or social status. You should expect that you will need other valued qualities to earn the highly attractive mates that you desire. Other important mating assets are included in the rest of MATING.

You can expect to attract many potential mates in an open setting. In fact, if you are extremely attractive you might even anticipate rather frenzied attempts by others to trade their non-appearance assets for your attractiveness. This is, of course, the kind of problem most people would love to have, but it is important for you to make sound decisions in the middle of all the excitement. Turn to page 167.

MATED APPEARANCE PROFILES

LLL Profile

The LLL Profile reflects your perception of parity in your present relationship. You see yourself and your mate as relatively unattractive and you consider appearance to be unimportant. The greatest danger with this profile, however, is that you might have acted out of resignation and a sense of inferiority when you selected each other. If you were offered greater attractiveness, would you take it? Both you and your mate are most likely to have the capability of being more attractive. Is it worth the effort to you?

In view of the research findings concerning the inordinate power of attractiveness in so many spheres of our society, would it not be worth the effort to work together to increase your appearance?

Another factor is important with regard to this LLL Profile: Can you truly be that unattractive or are you suffering from a damaged self-concept?

Keep in mind that attractiveness is largely a function of cleanliness and attention to grooming; you may be exaggerating your liabilities and overlooking strategies for improving your appearance.

Finally, the other mating factors which make up net worth are also important determinants of your happiness and success together and are discussed in detail in the rest of MATING. Turn to page 167.

LLM Profile

The LLM combination has the potential to create serious mating problems, unless special compensatory efforts are made. Putting it simply, you view yourself as low in appearance value and don't think appearance is important. However, your mate is of medium attractiveness. On this basis alone, you do not appear to have parity. Your mate has a higher value than you do. (Of course you gave it to your mate in your ratings, so it is not as if you didn't recognize the extra measure of attractiveness he or she has.)

You will need to have other major and continuing assets to balance against your mate's extra net worth in this factor. You also should review your own rating and consider whether it is appropriate or possible to improve your appearance. Keep in mind that much of attractiveness comes from effort and caring. Think again about the low value you have given to appearance, noting that because the majority of people rate it so high, it does have social value at least. In addition, when you do not attach much value to appearance, you effectively rob your mate of what would otherwise be a major asset in the relationship. Could this be an unconscious effort to maintain parity, or have you made this judgment based on a philosophical or religious position?

When you entered into the relationship with your mate you either had greater appearance value than you ascribe to yourself at present, or you compensated with other assets. It is important that you continue to review the dynamic of your relationship to maintain approximate parity. Turn to page 167.

LLH Profile

The appearance factor in your relationship is so unbalanced that you should anticipate serious mating problems. This is an unusual and dangerous profile.

The implications of the LLH combination are major. It requires major assets in other areas to exchange for high physical attractiveness. Of course, it could be that your mate has special liabilities which tend to balance out the asset of attractiveness. Usually when low attractiveness people mate, the low attractiveness person places a high value on the appearance of the mate and is willing to trade the assets of an extremely high net worth to gain the mate.

In your case, special circumstances apparently have produced an unusual relationship. Whether you and your mate selected each other because of other highly valued assets, or whether you have misjudged either your own or your mate's attractiveness, you must be particularly alert to problems. If you deny the importance of your mate's attractiveness, you are essentially robbing him or her of what otherwise would be a significant asset. Your motives for this should be reviewed to determine whether they flow from a philosophical viewpoint or an unconscious desire to keep your mate at your level. Turn to page 167.

LML Profile

This is a fairly common combination of scores and it reflects perceived parity. When individuals view themselves as relatively unattractive, they most often mate with others who are similar to them.

The complication here is that your medium interest in appearance could translate into some dissatisfaction with your mate's low attractiveness. On the other hand, your concern with appearance can be taken as a sign of good mental health if you are motivated to resolve what could be a difficult problem for your relationship.

Almost everyone can improve a low appearance score by concentrating on such appearance factors as grooming, clothing, complexion and posture, not to mention weight. If you and your mate have parity now, it is best for both of you to improve appearance at the same time. You might be surprised at the number of mating difficulties caused by one partner suddenly going on a self-improvement binge. It should be noted that the complementary and joint effort to make improvements in the net worth factors is at the very heart of the Dynamic Mating System. Turn to page 167.

LMM Profile

This is a fairly frequently encountered profile. You consider yourself to be relatively low in physical appeal, but think that appearance is of medium importance and have found a mate with medium appearance to match your interest.

With reference to appearance alone, you and your mate do not have parity, so you will need to continue to provide other compensating assets to achieve parity in net worth. Other mating factors and strategies for increasing your net worth are included in the rest of MATING.

Low appearance individuals with healthy self-concepts tend to be drawn to others with somewhat greater attractiveness. This is apparently what has happened to you. Unless low attractiveness individuals possess exceptional net worth, however, they generally avoid competing in the rarified strata of the high appearance crowd.

Keep in mind that your attractiveness almost certainly can be improved with effort, and the increased attractiveness pays astounding dividends. Turn to page 167.

LMH Profile

This is a fairly rare profile which is found among high achievers. Either you are extremely generous in your judgment of your mate's appearance, or you have some very high net worth assets to fair trade for your mate's attractiveness. With your low self rating for appearance but medium value placed on it, it is likely that you are an extremely harsh critic of yourself and are something of a perfectionist.

The essential issue is to determine whether those compensating assets still result in perceived parity between you and your mate. If they do not, then the relationship will reflect the strains associated with non-parity mating. Keep in mind that attractiveness almost always can be improved with effort and such factors as grooming, posture, weight, and so forth make up most of what is considered attractiveness. Turn to page 167.

LHL Profile

Help! The LHL Profile represents a very high risk for you and your relationship. Why? Because you see yourself as unattractive, you are extremely interested in having a mate who is high in appearance, but you see your mate in the same way you see yourself, as unattractive.

In one sense, it is not surprising that you have rated yourself and your mate as approximately equal in attractiveness. It represents parity, after all. However, this profile suggests much more. It also indicates that you really want an attractive mate but have been unable to find one. What to do?

One possibility is for both of you to give some substantial time and effort to the task of improving your appearance. As you indicate with your high interest in the appearance factor, it is worth it. Remember that such factors as clothing, grooming, weight and posture, which make up most of what is considered attractiveness, can be improved dramatically. Turn to page 167.

LHM Profile

The LHM Profile can represent mating problems, but also has positive implications. You view your own appearance in a negative light but are very interested in having a mate with excellent appearance. Your mate is seen as being of medium attractiveness.

It is apparent that your low self rating has not limited your aspirations. Sometimes this rating combination indicates a tendency toward being a perfectionist, so that the low rating for self and then medium rating for mate are somewhat more critical than they might otherwise be. With appearance as important as it is to you, you need not be reminded that such appearance factors as posture, grooming, clothing and weight can be managed for greater attractiveness.

Keep in mind that to achieve parity with your mate, you will need other non-appearance assets or your mate will be undercompensated in the relationship. Turn to page 167.

LHH Profile

This is the profile of a successful individual who truly cares about appearance and has the assets to find an attractive mate. The LHH Profile is relatively successful, providing the mate with the low rating has an abundance of other assets to fair trade.

In considering your mating relationship, ask yourself whether the parity continues. Are the assets which first attracted you to your mate, and vice-versa, still intact?

The danger with this combination is that a single-minded emphasis on appearance can place you at the disadvantage of being an essentially unequal partner who must take herculean efforts to sustain approximate parity. If your other assets, usually status and wealth, become diminished, your mate is left in an undercompensated position. Let's hope your other assets continue to grow, for this profile has typified the exchange of many highly successful couples through the centuries. Turn to page 167.

MML Profile

There is much potential for difficulty with this combination of ratings. You view yourself as moderately attractive, have a medium interest in the attractiveness of your mate, but you consider your mate to be relatively unattractive. You need to answer several important questions.

How much less attractive is your mate then yourself? How important is it that your mate is less attractive than you would prefer? What other compensating assets does your mate have to arrive at parity with you?

Your medium interest in your mate's appearance suggests that you would prefer it if he or she were more on a par with you.

Keep in mind that of the mating factors treated in this book, none is more easily changed than appearance, since most of it is made up of variables which can be improved readily, such as grooming, posture, weight, clothing and so forth. It is essential to work on improvements in appearance or other mating factors together, with mutual support and sensitivity. This is one of the key emphases of the Dynamic Mating System. Turn to page 167.

MMM Profile

All things being equal, this perfectly balanced profile should not pose relationship problems. From your standpoint, you have appearance parity with your mate. This is not to say that you cannot or should not attempt to improve in any way you deem feasible, including in attractiveness. In fact, your medium interest in appearance can be interpreted as indicating a moderate desire to make improvements. Keep in mind that attractiveness is readily increased with attention to weight, grooming, posture, clothing and so forth. Turn to page 167.

MMH Profile

From the point of view of mating success, the MMH combination is a positive one. You see yourself as of medium appearance, you have a medium amount of interest in your mate's appearance and you view your mate as very attractive.

There are two dangers to keep in mind with this profile. One is that your appearance is not as attractive as your mate's, which requires that you make up the deficit with other non-appearance assets.

The other danger is that your mate's attractiveness exceeds your interest in appearance, so that you will not value your mate's appearance sufficiently to give full credit for this asset which is so highly valued by almost everyone. If you do not give credit, you may have a distorted view of your mate's actual net worth and this, of course, will affect your parity calculation.

If you give some attention to these possible dangers, which are not insurmountable, the positive aspects of having an attractive mate can add much to your relationship. Remember that you can add to your own level of attractiveness by concentrating on the appearance

variables which can be managed readily, such as complexion, weight, muscle tone, grooming, and the like.
Turn to page 167.

MLL Profile

This is the profile of a person who has little interest in the attractiveness of a mate, but is more concerned about the mate's other net worth assets. Because you have greater appearance value than does your mate, you may feel you are not being fairly compensated. If your partner has other assets which balance your own net worth, then parity will be sustained.

It is unusual for a person with medium attractiveness to be so little interested in appearance. Most people who place a low value on a mate's appearance also see themselves as lacking attractiveness. There is a possibility that you may have a relatively poor self-concept and have not given yourself the credit that most people do for being attractive. There is also the possibility that your mate has influenced your thinking about attractiveness to the extent that you have rejected it as a valuable resource. Turn to page 167.

MLM Profile

The MLM Profile can best be summarized as reflecting appearance parity, but also indicating little interest in this mating factor. Some people with this combination feel that appearance deserves little attention. The danger is that little time or energy will be invested in maintaining attractiveness, and much of it will be diminished.

This is a relatively rare profile, inasmuch as most people with high or medium attractiveness also place a fairly high value on it. Some religious groups deliberately give attractiveness little value, but keep in mind that the majority of people have accorded a great deal of importance to it. We urge you to discuss the implications of this usually highly valued factor with your mate and be sure to find out whether attractiveness is given a high, medium, or low rating. You have appearance parity, but planning and thoughtful analysis are always appropriate. Turn to page 167.

MLH Profile

Your appearance profile has the potential to create major problems in your mating relationship. You see your own appearance as medium, you have little interest in your mate's appearance and yet you rate your mate as highly attractive.

The problem is the low value you attach to appearance. It is a virtual certainty that your mate's high appearance value is factored into the mating equation as an asset, but you have not given it much power. When you attach so little importance to appearance, it robs your mate of a quality usually given much value.

You may well have other net worth assets which equal the value of your mate's appearance, but it is unusual to trade them for high attractiveness when it is not considered important. It is possible that your mate will be tempted to become involved with others who place a high value on attractiveness.

Another possibility is that you are unconsciously attempting to keep the two of you equal by downgrading your mate's appearance advantage. This can be extremely dangerous and counterproductive, because even if you locked the mate in a closet, ultimately he or she will realize that the attractiveness factor is highly valued in the outside world. Turn to page 167.

MHL Profile

The MHL Profile is a dangerous one, in that it carries considerable risk for your relationship. You see your appearance as medium, you attach great importance to your mate's attractiveness, but have rated your mate as unattractive. The discrepancy between what you want and what your mate has suggests serious dissatisfaction or disappointment and you need to give careful consideration to strategies for reducing the disparity.

Frequently this profile is found when two fairly attractive people, who have parity, marry and over the course of years one becomes less attractive, usually by gaining weight or paying little attention to grooming.

Because you value attractiveness so much, it must be especially frustrating to you and to your mate to have such an appearance discrepancy. He or she may have other assets which make up for the imbalance, but of all the mating factors, appearance is one of the easiest to improve. The Dynamic Mating System, and perhaps some professional consultation, can be very helpful to you in this predicament. Turn to page 167.

MHM Profile

You have parity in your appearance, but you place a high value on attractiveness. Although both of you have medium appearance, you may be somewhat dissatisfied with the status quo and be eager for improvement for both of you. The appearance factor is one of the easiest of the mating factors to improve, particularly because so much of appearance is a function of grooming, clothing, weight, posture, and so forth.

In the give-and-take of a successful relationship there are many compromises. You may decide that your mate's other qualities are sufficient to compensate for his or her less-than-ideal appearance, or you may choose to work together toward improved attractiveness. In either case, it is important to factor in the other mating variables, and the Dynamic Mating System in the rest of the chapters can be beneficial to you in this effort. Turn to page 167.

MHH Profile

It is apparent from this profile that you have what you want, a very attractive mate. The potential problem is whether you have other net worth assets to trade for your mate's superior physical attributes. Does your mate also place a high value on appearance, and if so, does he or she seem satisfied with your level of attractiveness?

From most standpoints, this MHH Profile is a positive one, because you recognize the significance of attractiveness and appreciate it in your mate. To the extent that you can increase your own attractiveness, through grooming, weight control, exercise, and so forth, you probably will be more satisfied with the relationship. Turn to page 167.

HLL Profile

This is typically the profile of a person who has relied heavily on the Fair Trade Principle. You are likely to have traded your superior attractiveness for your mate's powerful non-appearance assets. This has not extracted a high price from you, because you have relatively little interest in your mate's appearance anyway.

However, unless your mate has other significant net worth assets, your relationship has not achieved parity

and could be in for some serious difficulty. You will
need to consider this possibility carefully.

One has to wonder why you have rated appearance as
of little importance when you are attractive yourself.
This is a fairly rare combination, sometimes reflecting a
lack of awareness of how influential appearance is, and
sometimes indicative of a well-thought-out philosophical point of view. Turn to page 167.

HLM Profile

You have greater appearance assets than your mate,
but this does not necessarily indicate an inequitable relationship. Your profile shows that you are attractive, you
care very little about your mate's appearance, and your
mate is of medium appearance. This is a fairly unusual
combination, in that most people who are attractive also
place a high value on appearance. You may have decided that, in the greater scheme of things, appearance
is not important to you, or you may somehow be unaware of how highly valued it is in our society.

Many people with the HLM Profile have fair traded
their attractiveness for the other valuable net worth assets of their mates. The two factors traded for most
frequently are economic advantages and social status,
although sometimes personality, or sensuality factors
are valued highly enough that attractiveness is traded
for them.

It is important in your relationship to maintain parity.
Sometimes mates who have less attractiveness will try to
convince the one with greater attractiveness that appearance is not important. You should be aware that
this maintains the appearance of parity, to the advantage of the less attractive mate, particularly when the
mate has few other assets to trade. Turn to page 167.

HLH Profile

This is an extremely rare profile. You see yourself and your mate as being very attractive but attach little significance or value to appearance. You should refer back to the appearance inventory and review your responses. Did you really mean to say that appearance is not important to you?

Sometimes attractive people, like people of all appearance levels, decide that appearance is, to them, not an important variable. This may come from a philosophical/religious or zealously political point of view, or you may have been around so many attractive people that is has ceased to be as significant to you as it is to almost everyone else. You might want to keep in mind that you both have an extremely important asset if you care to use it, whether for your own gain or the benefit of society. Turn to page 167.

HML Profile

The HML Profile is fairly rare, usually found among the upwardly mobile. You see yourself as high in attractiveness, have a medium interest in appearance, but see your mate as unattractive.

You should be aware of the significant disparity between yourself and your mate in terms of appearance value. Parity requires that your mate have some other valuable net worth assets to fair trade for your attractiveness. Many people with this profile have elected to trade for economic advantages or higher social status. This kind of fair trade can permit a stable and lasting relationship. Keep in mind, however, that if either attractiveness or the balancing assets are diminished, the relationship can become unequal very quickly. You may be at a serious disadvantage if most of your net worth is tied up in appearance. Consider the value of a diversified approach, and develop other assets. Turn to page 167.

HMM Profile

This is not an uncommon combination, particularly for women who have traded their attractiveness for the greater earning power and other advantages held by men. Whether you are male or female, however, you likely have exchanged for economic or social assets possessed by your mate, with such factors as sensuality and personality usually less important to the trade.

It should be recognized that, based on appearance alone, you do not have parity. The disparity is not too great to deal with, however, and in general, your profile is a positive one. You may want to develop other net worth assets even further, to avoid too much dependence on attractiveness alone. The Dynamic Mating System should be helpful to you in this effort. Turn to page 167.

HMH Profile

Your relationship has the advantage of two attractive people, and with the importance attached to appearance by most of the population, the advantage is significant indeed. However, you do not seem to consider appearance to be more than of medium importance, perhaps because you presently have it and, to a certain extent, take it for granted. It is unlikely, however, that mere chance resulted in your finding each other. The relationship reflects parity in appearance, and no special problems are apparent. Keep in mind that with the appearance combination you have you can use its considerable leverage to great advantage in your relationship or with people in general. Turn to page 167.

HHL Profile

This is one of the most dangerous profiles for a relationship. You view yourself as attractive, consider appearance to be extremely important, but have an unattractive mate. This profile usually is found among those who exchanged their own attractiveness for the social or economic advantages of their mates. One danger is that you may lose some of your attractiveness advantage and risk losing parity. Another is that you may become discouraged with your mate's relative unattractiveness and decide the trade is no longer fair. In addition, if your mate loses some of the other assets, he or she will no longer have parity, and you are likely to be more attracted to other potential mates with the attractiveness you value so highly.

It is possible that you are something of a perfectionist, unable to appreciate the attractiveness that your mate does have. You also might consider ways in which your mate could improve in appearance, including weight, better grooming and the like. Turn to page 167.

HHM Profile

There is some potential for mating problems with this profile, although nothing which cannot be overcome with thought and effort. Although the HHM combination does not reflect a parity relationship, the discrepancy is not overly great and can be made up with other assets. The greater danger is seen in the difference between what you desire in a mate's appearance and your mate's failure to reach that level.

This combination is seen most often among those who began with a relationship which had parity, but with time, one of the partners either improved significantly in appearance or lost much of his/her attractiveness.

Weight changes often contribute to appearance changes, for example.

You will learn much more about your relationship as you progress through MATING, but the lack of perceived parity in appearance can be a disruptive force. Give it some consideration soon. Turn to page 167.

HHH Profile

What a team! Both of you have the advantages of being attractive and the awareness of how important this attribute can be from the standpoint of helping you to control your own destiny.

The HHH combination is based on parity, and the one limitation which is sometimes seen is that the parties will emphasize appearance to the relative exclusion of the other net worth variables. With the advantages of excellent appearance and the motivation to maintain this edge, you have a splendid platform from which to develop other assets, to be even more successful in your life together. Turn to page 167.

8

Money and Mating

THE MONEY SIDE of mating has always been an important factor. Take, for example, the price of wives and husbands. You are aware of the practice of wife-buying throughout history. Recently in New Guinea, for example, the going rate that a husband had to pay a virgin bride's father was five pigs, one cassowary, and $300. Of course, if the bride had been married before, the price dropped to two pigs, one cassowary, and $37.50. Now that's depreciation!

The use of the dowry was actually a thinly disguised strategy for purchasing a husband. The dealing and wheeling back in the dowry days must have been something to hear:

> "Listen, you look like a nice young man. I'll tell you what. Not only will I throw in the lower 40 hectares and three ewes, how would you like some linen? Just feel the goods, kid. So whaddaya say? Have we got a deal? What? Of course she's a virgin! Tell you what . . . I'll throw in a cassowary."

We don't make these kinds of deals all that often in America, of course, but economic factors are still important. In fact, so many marriages are planned with a pooling of resources in mind that one has to wonder if the object is matrimony or merger!

When you complete the tests in this chapter, we shall see how important the economic factor is to your mating decisions.

Note Carefully Whether the Tests and Instructions That Follow are for Unmated or Mated.

ECONOMIC STATUS INVENTORIES

Test 1

PERSONAL ECONOMIC STATUS
Unmated and Mated

Directions: This inventory is to be completed by both unmated and mated respondents. Rate *yourself* from 1 to 10 with regard to the following 15 statements. A rating of 1 means the statement is not at all like you, 5 is average, and 10 means the statement describes you almost perfectly.

1. ______ I have access to enough money to live well without major concerns for the future.
2. ______ I have friends and family who would be helpful financially if I needed help.
3. ______ I earn a good salary.
4. ______ I am likely to inherit major amounts of money.
5. ______ I have access to major amounts of credit if I need it.
6. ______ People see me as prosperous.
7. ______ My education and skills give me economic security.
8. ______ I have a good investment sense.
9. ______ I have a secure job and career.
10. ______ I have the "Midas touch". . . things work out well for me economically.
11. ______ I manage money well.
12. ______ If I needed to, I could find a different job paying a comparable amount.
13. ______ I have a retirement plan, health insurance, etc.

14. _______   I have a great future in my work.
15. _______   I know how to make money go a long way.
    _______   TOTAL

(If unmated, proceed to Test 2. If you have a mate, go directly to Test 3.)

Test 2

THE ECONOMIC FILTER
Unmated Only

Directions: The purpose of this test is to determine the level at which you filter out potential mates who fail to meet your minimum economic standards. Which is the *lowest* economic level acceptable to you in a mate?

1. _______ Low   The mate I select can be poor if other fair-trade assets are sufficient.

2. _______ Medium   I would not select a mate, irrespective of other assets, with less than medium economic status.

3. _______ High   Notwithstanding any other assets a potential mate might have, I will only accept a mate with major economic resources.

(Unmated go directly to Test 3.)

Test 3

IMPORTANCE OF ECONOMIC STATUS
Unmated and Mated

Directions: Both Unmated and Mated answer Question A. Only Mated respond to Question B.

A. How much importance do you attach to economic status, as reflected by motivation to improve it? (Check the appropriate level.)

Low _______ Medium _______ High _______

(Levels are described below.)

B. How much importance does *your mate* seem to attach to economic status, as reflected by apparent motivation to improve it. (Check the appropriate level.)

Low _______ Medium _______ High _______

Low: Economic status is not important and it's not worth making an effort to improve it.

Medium: Economic status is important and it's worth making a reasonable effort to improve it.

High: Economic considerations are of great importance and a major effort to improve economic status is clearly worthwhile.

(Unmated turn to Scoring Instructions on page 173. Mated proceed to Test 4.)

Test 4

MATE'S ECONOMIC STATUS
Mated Only

Directions: This test examines your views of your mate's economic status. Respond to each item as it relates specifically to your mate. DO NOT refer to responses on earlier tests. Rate your *mate* using a scale of 1 to 10, with 1 being the lowest, 5 average and 10 highest.

1. _______ My mate has access to enough money to live well without major concern for the future.

2. _______ My mate has friends and family who would be helpful financially if she/he needed help.

3. _______ My mate earns a good salary.

4. _______ My mate is likely to inherit major amounts of money.

5. _______ My mate has access to major amounts of credit if it is needed.

6. _______ People see my mate as prosperous.

7. _______ My mate's education and skills give him/her economic security.

8. _______ My mate has a good investment sense.

9. _______ My mate has a secure job and career.

10. _______ My mate has the "Midas touch"... things work out well economically for her/him.

11. _______ My mate manages money well.

12. _______ If my mate needed to, he/she could find a different job paying a comparable amount.

13. _______ My mate has a retirement plan, health insurance, etc.

14. ______ My mate has a great future in her/his work.

15. ______ My mate knows how to make money go a long way.

______ TOTAL

(Mated turn to Mated Scoring Instructions on page 175.)

Scoring Instructions

ECONOMIC INVENTORIES
Unmated Only

Directions:

Step 1. Add the numbers in the Score Column in Test 1 (Economic Status) on page 169.

Place the Total Score here: ________ .

Step 2. Using the conversion chart below, convert your Total Score from Step 1 to Scaled Points and place a (√) by the appropriate level:

Conversion Chart

Score from Step 1		*My Scaled Points*	(√)
15 to 41	=	10 Points	________
42 to 68	=	20 Points	________
69 to 95	=	30 Points	________
96 to 122	=	40 Points	________
123 to 150	=	50 Points	________

Step 3. Look back at Test 3A (Importance of Economic Status) on page 171.

What was your rating? _________________
(Low, Medium, or High)

Convert this rating to a number by using the following scale and circle the appropriate number below:

Low = 1   Medium = 2   High = 3

Step 4. Multiply your Scaled Points (from Step 2 above) by 1, 2, or 3, depending on which number you circled in Step 3.

For example, if you have 30 Scaled Points in Step 2 and your level of importance score from Step 3 is Medium (2), then multiply 30 × 2 = 60.

Do your computations and then write your Economic Status Score here:

Economic Status __________ .

Step 5. In the conversion chart below, you will see how your Economic Status Score rates in the mating marketplace. Place a (√) in the appropriate level.

(√)   *Economic Status Level*

__________   10 to   50 = Low Economic Status
__________   60 to 100 = Medium Economic Status
__________   110 to 150 = High Economic Status

(Unmated now turn to the text of Chapter 8, starting on page 180.)

Scoring Instructions

ECONOMIC INVENTORIES
Mated Only

Directions:

Step 1. Add the numbers in the Score Column in Test 1 (Economic Status) on page 169.

Place the Total Score here: __________

Step 2. Using the conversion chart below, convert your Total Score from Step 1 to Scaled Points and place a (√) by the appropriate level:

Conversion Chart

Score from Step 1		*My Scaled Points*	(√)
15 to 41	=	10 Points	__________
42 to 68	=	20 Points	__________
69 to 95	=	30 Points	__________
96 to 122	=	40 Points	__________
123 to 150	=	50 Points	__________

Step 3. Add the numbers in the Score Column in Test 4 (Mate's Economic Status) on page 173.

Place the Total Score here: __________

Step 4. Using the conversion chart below, convert your *Mate's* Total Score from Step 3 to Scaled Points and place a (√) by the appropriate level:

Conversion Chart

Score from Step 3		Mate's Scaled Points	(√)
15 to 41	=	10 Points	_________
42 to 68	=	20 Points	_________
69 to 95	=	30 Points	_________
96 to 122	=	40 Points	_________
123 to 150	=	50 Points	_________

Step 5. Look back to Test 3A (Importance of Economic Status) on page 171.

What was your rating? _________________
(Low, Medium or High)

Convert this rating to a number by using the following scale, and circle the appropriate number below:

Low = 1   Medium = 2   High = 3

Step 6. Return to Test 3B on page 171. What was your *Mate's* rating?

Mate's rating: _________________
(Low, Medium or High)

Convert this rating to a number by using the following scale, and circle the appropriate number below:

Low = 1   Medium = 2   High = 3

Step 7. Next we will compute Your Economic Status. Multiply *My* Scaled Points (Step 2) by 1, 2, or 3, depending on which number you circled in Step 5.

For example, if you have 30 Scaled Points from Step 2 and you have circled 2 (Medium) in Step 5, then multiply 30 × 2 = 60S. We have placed an "S" next to the

score. It stands for "Self," as this is your own Economic Score.

Do your computations and write your Economic Status Score here:

My Economic Status Score ＿＿＿＿＿ S (Self)

Step 8. Now we need to compute your *Mate's* Economic Status Score. Multiply your *Mate's* Scaled Points (Step 4) by 1, 2, or 3, depending on the Importance Rating circled in Step 6.

For example, if your mate has 20 Scaled Points (in Step 4) and the Level of Importance Score (from Step 6) is 2 (Medium), then multiply $20 \times 2 = 40$M. We have included an "M" next to the score. It stands for "Mate," as this is your *Mate's* Economic Status Score.

Now do your computations and write your *Mate's* Economic Status Score here:

Mate's Economic Status Score: ＿＿＿＿＿M (Mate)

Step 9. The conversion chart below translates the Economic Status Score for you and your mate into levels for easy comparison.

Place an "S" (Self) by the appropriate level for *your* score from Step 7. Place an "M" (Mate) by the appropriate level for your *Mate's* score from Step 8.

"S" or "M"	Economic Status Level	
＿＿＿＿＿	10 to 50	Low Economic Status
＿＿＿＿＿	60 to 100	Medium Economic Status
＿＿＿＿＿	110 to 150	High Economic Status

Step 10. Look at your Economic Status Score in Step 7 and your *Mate's* Economic Status Score in Step 8. Which is larger? To find your Parity Score for Economic Status, subtract the smaller number from the larger number and carry to your answer the "S" or "M" from the larger number.

For example, if your Economic Status Score is 90S and your *Mate's* Economic Status Score is 50M, your Parity Score for Economic Status is 40S. (See example below.)

Example:
$$\begin{array}{r} 90\text{S} \\ -\underline{50\text{M}} \\ 40\text{S} \end{array}$$

Write your Parity Score for Economic Status below:

Parity Score for Economic
Status __________ ("S" or "M")

Step 11. The more similar you and your Mate's Economic Status Scores, the smaller will be your Parity Score. Perfect parity has a score of zero. The range of possible parity scores is from 0 to 140S or 0 to 140M. Remember, if the Parity Score ends with an "S," this means you have greater power with the Economic Status Factor than your mate. If it ends with an "M," your mate has greater power. The chart below will be helpful in interpreting your Parity Score.

Parity Score Guide

"S" (Self) Scores

Zero	Perfect parity.
10S to 40S	Excellent balance in economic power.
50S to 80S	Medium balance, but you bring greater economic power to the relationship than your mate.
90S to 140S	Dangerously unbalanced, unless your relationship includes compensating fair-trade assets. You have much greater economic power than your mate.

"M" (Mate) Scores

Zero	Perfect parity.
10M to 40M	Excellent balance in economic power.
50M to 80M	Medium balance, but your mate brings greater economic power to the relationship than you.
90M to 140M	Dangerously unbalanced, unless your relationship includes compensating fair-trade assets. Your mate has much greater economic power than you.

(Now turn to the text of Chapter 8 on the next page.)

Money and Successful Marriage

Three main points should be made at the outset:

(1) The economic factor is extremely important to mating success.
(2) You do not have to have high economic power to enjoy marital success and happiness . . . but it certainly helps.
(3) The key is, as always, parity.

The research evidence connecting the economic factor and marriage success is both consistent and compelling. Here is a sample of some of the findings:

- Couples who enjoy higher incomes have lower divorce rates.[1]
- Long periods of unemployment lead to high divorce rates.[2]
- Money by far is the greatest area of disagreement between spouses.[3]
- The economic factor is among the top three causes of serious marital problems.[4,5,6]

There is no doubt about it, the economic variable is a powerful force in mate selection and marriage success. Actually, it probably shouldn't surprise anyone too

---

What Price Freedom?

The largest divorce settlement on record occurred in California in 1979. Barbara Cooke divided a $200,000,000 fortune with her sports promoter husband, Jack Kent Cooke. Under California's equal property law, that means each kept one hundred million dollars.[13a]

much, considering how financial success seems to contribute to success in almost all other facets of life.

The phenomenon of success begetting success, which we described with regard to personality, operates with the economic factor too. Those of you who achieved high economic scores probably are aware of how other benefits follow. For example, people with high economic power can use that mating asset to attract others with wealth. If they marry, they will pool their already abundant resources.

It's something like the banks voluntarily sending credit cards and extending lines of credit to the super-solvent people who need them least!

Most people put economic factors right at the top of their list of "Things to think about before getting married."

Even in the giddy midst of euphoric love, they are not likely to forget about the financial side of mating. While they are declaring, "We'll live on the nectar of love, my sweet!" they are likely to be doing some mental arithmetic and comparing financial assets and liabilities. As one client told us,

> "Even when we were promising the moon to each other, we were both trying to figure out how we could pay for a place to live and still enjoy the luxury of eating."

How much economic power do most people want in a mate?

While most say they wouldn't mind terribly if they found someone with riches to spare, in actuality they really want what research tells us they should have. What it is? Here's a hint: It's what we're going to discuss next.

Economic Parity

A client recently was describing the kind of mate she

was looking for and when she got to financial characteristics, she said something like this:

> "I'll tell you, though. I'm not really sure I want to
> get married. Too many of the guys I know waste
> their money, and haven't any savings at all. I have a
> different attitude and I save my money. I've got
> $30,000 in an account and I own my car free and
> clear. Why should I get married and have to share
> my money with anyone? I worked too hard for it."

Selfishness? Not really. It's more of a concern about
parity, and it's justifiable.

Financial parity in marriage includes approximate
equality in earning power and savings. To the extent
that the partners contribute equal amounts to a marriage, their relationship will benefit. They will appreciate one another's contributions and the resent-
ments and misunderstandings of non-parity will be
avoided.

Women are at a serious economic disadvantage, of
course, so they are not likely to bring as much economic
power to a marriage as will men. Men have enjoyed a
financial advantage throughout history, and the recent
gains by women have only slightly narrowed the money
gap.

In a strict sense, most marriages are built on economic
non-parity. However, if either one of the partners fore-
goes or discontinues employment in order to benefit
the relationship, it is crucial that he or she be credited
with that important contribution. Otherwise, the non-
parity created will be too destructive to overcome. This
is illustrated in the case of Lorna and Roger.

LORNA CLEANS WHILE ROGER PLAYS

When Lorna and Roger were married a year after
high school, they had a lot in common. They had the

same friends, enjoyed dancing and sports, were nice looking, and came from the same neighborhood. They didn't have much money, but they had the same dreams.

Twenty-six years and three grown children later, Roger and Lorna were wealthy and, in the eyes of most people, successful. Roger owned a thriving construction business, they had purchased a huge home in the most exclusive area of the city, and their children were in college or professional school. Unfortunately, their marriage was in deep trouble.

What had gone wrong?

They had many areas of minor contention, but their biggest problem was money. It was not that they had too little income or savings. They had too little parity.

They had married on equal terms, and while Lorna was raising the children, they remained equal. Roger was responsible for supporting the family and she was responsible for raising the kids. As long as she had that job, which they both viewed as extremely important, they continued to have parity. When the children were grown and flown, however, she was out of work.

Roger was eager for some adventures. "Off to Tahiti!" he cried. "Let's head for weekends in Las Vegas." But Lorna was in no mood for such frivolity. She was out of a job. Her career was over. She had no other marketable skills, and after being so frugal for so long, and letting Roger make all the financial decisions, she felt like a kept woman. She had lost her purpose, her status, her pride, and as people do when they suffer a loss, she became depressed.

She began to pour more and more time into the house, and she let the cleaning woman and the lawn service go. She became more thrifty than ever, regressing back to the early days when she made curtains out of leftovers and worked from dawn to dusk with her wifely chores. Roger was disgusted at this display and increasingly spent time away from home, off on a variety of trips to "enjoy what I've worked so hard to have."

Lorna became more anxious and depressed than ever. She felt that Roger was more attractive than she, had more social skills and worlds of confidence, and he earned all the money. What could she do? She thought of taking a job as an office worker, but had no skills. Instead, she continued to try to *save*, through frugality and hard work, as much money as Roger earned. It was, of course, a lost cause. She could not hope to compete.

On one of his many flights to adventure, Roger met a pretty, young airline attendant who was attracted by his immaculate grooming, his confidence, his sensuality, and last but not least, his money. She had net worth assets for Roger that approached parity, and Lorna's value had been reduced to that of a superb cleaning woman.

The marriage ended a short time later.

What happened to Lorna? Don't feel too sorry for her. After several months of counseling and with the support and encouragement of friends, one of whom was a creative banker, she began a small business of her own and poured all of her energy and time (and anger) into it. She became very successful almost overnight. It's just too bad she had not been involved as a partner with Roger's business years before, so that their parity could have been protected.

Women and Money

You are aware of the strides women have made in coming closer to economic parity with men, but how did they ever get in such a hole in the first place? Why have women been willing to trade on attractiveness for economic gain, for example? And why is it that they still earn only 60% of what men earn?[7]

A very brief history lesson is in order. For thousands of years, women have been considered the property of their fathers or husbands. In the Greek era, a thousand years before Christ, women were considered chattel and could be traded or lent.[8] A thousand years later in Rome, they could not enter any legal contract, own property, or even inherit the family home.[9] Another thousand years later, in medieval days, a woman still could not transact business or inherit land.[10]

The industrial revolution finally brought some changes. Women were given the "privilege" of working outside the home, in mills and factories. They were paid less than half of what males were paid for equivalent work, however.[11]

In light of this background, it is not difficult to understand the importance women have attached to marrying a man with wealth. The choice was, and often remains, between working long hours in relative poverty, unable to provide adequately for their children with little hope for the future; or having the comforts of a home, perhaps some luxuries, and much brighter prospects for the children. Viewed in this perspective, how important was a handsome face or muscular chest? How important was love? For many women the logical and intelligent thing to do was to trade their appearance assets for economic advantage. Given the circumstances, who can fault them?

Sex Roles and Marital Success

Central to any discussion of money and marriage is a larger issue: marriage roles. Who does what? Who has control over which aspects of married life? Who makes the decisions for the family, including the all-important financial decisions?

We see these issues translated every day into the behaviors of married life. In a traditional male dominated relationship, the wife cleans the house, cooks the meals, tends the children, and looks after the day-to-day needs of the family. The husband on the other hand, functions as the CEO (Chief Executive Officer). He allocates responsibilities, makes the money, and sets the framework within which it will be spent or saved, although the wife may actually pay the bills and do the shopping. The wife may have outside employment if the husband consents, but her domestic role and workload responsibilities usually remain as great as ever!

Reacting to the many inequities of traditional married life, some people seem to believe that each mate should share equally in each task and responsibility. The wife should help with the lawn and the husband with the dishes; the wife should face down the intruder and the husband sew the daughter's torn dress. But partners don't need to assume identical roles to have parity.

If the female prefers being a mother and housewife, and she and her husband view it as having parity with his contribution, then the relationship can be a healthy one. If the male would prefer being a "househusband," and he and his wife view it as having parity with her contribution, then this relationship can be satisfactory too.

The role combinations within a family are infinite. The two keys to role compatibility are (1) that power to determine the roles is shared jointly and equally, and (2) that each partner sees the roles as equitable.

Economic parity does not mean that both partners must hold outside employment, or that if they do, they need to earn identical salaries. It is essential to the health of a marriage that both partners recognize the value that each brings to the relationship and view the contributions as equal. If one is primarily responsible for managing the home and family, this contribution must be recognized for its genuine value. In addition, provisions must be made for modifying the economic system with the passage of time and changed circumstances.

Marriages prosper when both partners view the benefits and responsibilities as equitably distributed. Relationships deteriorate when either party feels undercompensated. No relationship can afford to be caught in the trap of economic non-parity.

Attitudinal Parity

Another contributing factor for economic parity is the *attitude* that both partners have toward money and how it should be spent, saved, or invested. We refer to this as attitudinal parity.

The most successful marriages are those in which the partners come from similar economic backgrounds.[12,13] This is the case largely because they share similar expectations and attitudes about money.

When people from different economic backgrounds marry, the ones who come from the lower level are likely to feel inadequate and inferior. They may appear extra thrifty or "penny wise and pound foolish" to the persons from higher levels. One reaction to the feelings of uncertainty they experience is to glamorize the virtues of the less wealthy life they left behind. They are apt to say such things as, "We didn't have much but we had love," and, "We knew how to have fun without money."

The persons from higher economic backgrounds also often experience uncertainty and dissatisfaction as the expected privileges they grew up with are no longer available.

Does this mean that people should not marry across socio-economic levels? No, but it does mean that the one who moves up should bring to the relationship other assets which boost his or her net worth up to parity with the mate. Otherwise, the marriage will suffer, as all inequitable relationships do.

Attitudes about money are usually quite different at the different economic levels, and different attitudes figure in most of the money arguments in marriage. In fact, recent research has found that disagreements concerning money are not so much about the amount of income, but how it should be handled.[14] Many couples endure endless battles about how money should be spent or saved, whether both spouses should work or not, whether savings accounts should be marked "his" and "hers," and who owns and is responsible for what. Once euphoric love has burned out, this very practical issue comes to the forefront of most marriages.

Your Economic Power

Whether you scored high, medium or low, you should be aware of the advantages of higher economic net worth. This is not to say that one cannot enjoy an excellent marriage relationship at a lower financial level, but it means beating the odds. The correlation between financial success and mating success is well documented.

By preparing yourself for a more financially rewarding career, you will be able to trade that economic advantage for other assets in your mate. Extra education and training, for example, pay off in many more ways than an increased salary. And don't forget the benefits of specialization we discussed in earlier chapters.

Whatever your economic lifestyle, be sure to look for parity in your mate, including similar plans and attitudes, and make the necessary effort to maintain parity through the changes that the passing years inevitably bring. Research is very clear on this point: Economic disparity corrupts mating relationships! If you are unmated, look again at the economic power you said would be the least you would accept (Test 2). If it is different from your own level, think about the need for parity and how you might achieve what you want. You can make the effort to increase your own level, or plan to fair trade other mating assets you might have for higher economic power in your mate. In any case, higher net worth levels in a mate have to be earned, or the relationship cannot prosper.

Finally, your motivation to increase your economic power should be matched by your mate's. Keep in mind how important it is to build a relationship based on mutual goals and expectations, including attitudes toward the money side of mating.

9

Sex and Affection

WHY HAVE WE COMBINED SEX and affection into one mating factor? Aren't they two separate behaviors?

They can be treated as separate, of course. Showing affection to someone is not necessarily being sexual. Conversely, sexual behavior often does not involve being affectionate. After all, sex can be a (dare we say it?) solitary activity, without any social and affectionate component at all.

But the mating factor we are especially concerned about is the *combination of shared physical and emotional satisfaction*. It is not sex by itself or the showing of affection to another by itself. It is the integrated combination of the two.

Listen to the kind of dialogue heard every day in the marriage therapy office and you'll have a better understanding of the sex and affection factor:

"There's something missing from our marriage."
"I'll say! And it was missing again last night! I'm
 getting tired of being elbowed in the ribs."

190

> "Well, if you'd come to bed earlier, maybe . . ."
> "Maybe nothing. You're just not interested."
> "What's the difference? You don't love me anyway!"
> "There you go again. I'm just not the kind of person who's all lovey-dovey and sweetie-sweet. It's just not my nature."

This is not a minor quarrel. One spouse is complaining about not *being* loved enough, and the other is upset about a shortage of love *making*.

When it comes to love in a relationship, however, it should not be a question of *being* versus *making*. As you are probably well aware, people need both. And the best way to get both is when they are provided in a combination of sex and affection.

The combination can include cuddling, intercourse, holding hands, hugging and kissing and other physical intimacies as well, combined with such affectionately loving phrases as:

> "You are so important to me."
> "What a wonderful person you are."
> "Did I ever tell you how I fell in love with you?"
> "I love your face."
> "I was thinking about you all day yesterday."
> "There's no one more exciting than you."
> "Guess what? . . . I love you."

We'll come back to the topic of sex and affection after you have completed the tests and calculated your scores. Have a good time.

Note Carefully Whether the Tests and Instructions That Follow are for Unmated or Mated.

SEX AND AFFECTION INVENTORIES

Test 1

SEX AND AFFECTION
Unmated and Mated

Directions: This test is to be completed by everyone, both unmated and mated. Rate *yourself* on each item on a scale of 1 to 10, with 1 being the lowest, 5 average, and 10 highest.

1. ______ I have a strong sex drive.
2. ______ I enjoy kissing and hugging my mate.
3. ______ I quickly become sexually stimulated.
4. ______ I like (would like) to give my mate sexual satisfaction.
5. ______ I enjoy showing affection.
6. ______ I enjoy conversations which concern love and feelings.
7. ______ I enjoy being shown affection.
8. ______ I prefer (would prefer) frequent sex.
9. ______ I am considered to be sexually desirable.
10. ______ I like cuddling.
11. ______ I enjoy (would enjoy) sexual variety.
12. ______ I prefer (would prefer) sex which lasts for a long time.
13. ______ I enjoy being hugged and kissed by my mate.
14. ______ I am sexually open and uninhibited.
15. ______ I have good feelings about my sexuality.
______ TOTAL

(If unmated, proceed to Test 2. If you have a mate, go directly to Test 3.)

Test 2

THE SEX AND AFFECTION FILTER
Unmated Only

Directions: The purpose of this test is to determine the level at which you filter out potential mates who fail to meet your minimum standards as it concerns affection and sexuality. Which is the *lowest* level below acceptable to you in a mate?

1. _______ Low   The mate I select can have little interest in sex or affectionate behavior if other fair-trade assets are sufficient.

2. _______ Medium   I would not select a mate, irrespective of other assets, with less than average interest in sex and affection.

3. _______ High   Notwithstanding any other assets a potential mate might have, I will only accept a mate who is highly affectionate and sexually enthusiastic.

(Unmated go directly to Test 3.)

Test 3

IMPORTANCE OF SEX AND AFFECTION
Unmated and Mated

Directions: Both Unmated and Mated answer Question
A. Only Mated respond to Question B.

A. How much importance do you attach to sex and
affection, as reflected in your motivation to
refine and improve the physical side of love?
(Check the appropriate level.)

Low _______ Medium _______ High _______

(Levels are described below.)

B. How much importance does *your mate* seem to
attach to sex and affection, as reflected in his
(her) motivation to refine and improve the
physical side of love? (Check the appropriate
level.)

Low _______ Medium _______ High _______

(Levels are described below.)

Low: Sex and affection are not very
important and its not worth making an
effort to improve them.

Medium: Sex and affection are important and it's
worth making a reasonable effort to
improve them.

High: Sex and affection are of great
importance and it's worth making a
major effort to sustain and even
improve this vital part of life.

*(Unmated proceed to Scoring Instructions on page 196. Mated
go to Test 4.)*

Test 4

MATE'S SEX AND AFFECTION
Mated Only

Directions: This test examines your views of your mate's attitude toward sex and affection. Answer each question as it relates specifically to your mate. Do NOT refer to responses on earlier tests. Rate your *mate* on the items using a scale of 1 to 10, with 1 being lowest, 5 average, and 10 highest.

1. ______ My mate has a strong sex drive.
2. ______ My mate enjoys kissing and hugging me.
3. ______ My mate quickly becomes sexually stimulated.
4. ______ It is important to my mate that I receive sexual satisfaction.
5. ______ My mate enjoys showing affection.
6. ______ My mate enjoys conversations which concern love and feelings.
7. ______ My mate enjoys being shown affection.
8. ______ My mate prefers frequent sex.
9. ______ I consider my mate to be sexually desirable.
10. ______ My mate likes cuddling.
11. ______ My mate enjoys sexual variety.
12. ______ My mate enjoys sex which lasts for a long time.
13. ______ My mate enjoys being hugged and kissed by me.
14. ______ My mate is sexually open and uninhibited.
15. ______ My mate has good feelings about his (her) sexuality.
______ TOTAL

(Mated go directly to Scoring Instructions on page 197.)

Scoring Instructions

SEX AND AFFECTION
Unmated Only

Directions:

Step 1. Add the numbers in the Score Column in Test 1 (Sex and Affection) on page 192.

Place the Total Score here: _________ .

Step 2. Using the conversion chart below, convert your Total Score from Step 1 to Scaled Points and place a (√) by the appropriate level:

Conversion Chart

Score from Step 1		*My Scaled Points*	(√)
15 to 41	=	10 Points	__________
42 to 68	=	20 Points	__________
69 to 95	=	30 Points	__________
96 to 122	=	40 Points	__________
123 to 150	=	50 Points	__________

Step 3. Look back at Test 3A (Importance of Sex and Affection) on page 194.

What was your rating?_______________________
(Low, Medium, or High)

Convert this rating to a number by using the following scale, and circle the appropriate number below:

Low = 1   Medium = 2   High = 3

Step 4. Multiply your Scaled Points (from Step 2 above) by 1, 2, or 3, depending on which number you circled in Step 3.

For example, if you have 30 Scaled Points in Step 2 and your level of importance score from Step 3 is Medium (2), then multiply 30 × 2 = 60.

Do your computations and then write your Sex and Affection Score here:

Sex and Affection Score __________ .

Step 5. In the conversion chart below, you will see how your Sex and Affection Score rates in the mating marketplace. Place a (√) in the appropriate level.

(√)	*Sex and Affection Level*		
__________	10 to 50	=	Low Sex and Affection
__________	60 to 100	=	Medium Sex and Affection
__________	110 to 150	=	High Sex and Affection

(Now turn to the text of Chapter 9, starting on page 203.)

Scoring Instructions

SEX AND AFFECTION INVENTORIES
Mated Only

Directions:

Step 1. Add the numbers in the Score Column in Test 1 (Sex and Affection) on page 192.

Place the Total Score here: __________ .

Step 2. Using the conversion chart below, convert your Total Score from Item 1 to Scaled Points and place a (√) by the appropriate level:

Conversion Chart

Score from Step 1		My Scaled Points	(✓)
15 to 41	=	10 Points	__________
42 to 68	=	20 Points	__________
69 to 95	=	30 Points	__________
96 to 122	=	40 Points	__________
123 to 150	=	50 Points	__________

Step 3. Add the numbers in the Score Column in Test 4 (Mate's Sex and Affection) on page 195.

Place the Total Score here: __________ .

Step 4. Using the conversion chart below, convert your *Mate's* Total Score from Step 3 to Scaled Points and place a (✓) by the appropriate level:

Conversion Chart

Score from Step 3		Mate's Scaled Points	(✓)
15 to 41	=	10 Points	__________
42 to 68	=	20 Points	__________
69 to 95	=	30 Points	__________
96 to 122	=	40 Points	__________
123 to 150	=	50 Points	__________

Step 5. Look back to Test 3A (Importance of Sex and Affection) on page 194.

What was your rating? __________________
(Low, Medium, or High)

Convert this rating to a number by using the following scale, and circle the appropriate number below:

Low = 1   Medium = 2   High = 3

Step 6. Return to Test 3B on page 194. What was your *Mate's* rating?

Mate's rating _______________________
(Low, Medium, or High)

Convert this rating to a number by using the following scale, and circle the appropriate number below:

Low = 1   Medium = 2   High = 3

Step 7. Next we will compute Your Sex and Affection Score. Multiply *My* Scaled Points (Step 2) by 1, 2, or 3, depending on which number you circled in Step 5.

For example, if you have 30 Scaled Points from Step 2 and you have circled 2 (Medium) in Step 5, then multiply 30 × 2 = 60S. We have placed an "S" next to the score. It stands for "Self," as this is your own Sex and Affection Score.

Do your computations and write your Sex and Affection Score here:

My Sex and Affection Score _________ S (Self)

Step 8. Now we need to compute your *Mate's* Sex and Affection Score. Multiply your *Mate's* Scaled Points (from Step 4) by 1, 2, or 3, depending on the Importance Rating circled in Step 6.

For example, if your mate has 20 Scaled Points (in Step 4) and the Level of Importance Score (from Step 6) is 2 (Medium), then multiply 20 × 2 = 40M. We have included an "M" next to the score. It stands for "Mate," as this is your *Mate's* Sex and Affection Score.

Now do your computations and write your *Mate's* Sex and Affection Score here:

Mate's Sex and Affection Score __________ M (Mate)

Step 9. The conversion chart below translates the Sex and Affection Scores for you and your mate into levels for easy comparison.

Place an "S" (Self) by the appropriate level for *your* score from Step 7. Place an "M" (Mate) by the appropriate level for your *Mate's* score from Step 8.

"S" or "M"   Sex and Affection Level

__________   10 to 50   Low Sex and Affection
__________   60 to 100   Medium Sex and Affection
__________   110 to 150   High Sex and Affection

Step 10. Look at your Sex and Affection Score in Step 7 and your *Mate's* Sex and Affection Score in Step 8. Which is larger? To find your Parity Score for Sex and Affection, subtract the smaller number from the larger number and carry to your answer the "S" or "M" from the larger number.

For example, if your Sex and Affection Score is 90S and your Mate's Sex and Affection Score is 50M, your Parity Score for Sex and Affection is 40S. (See example below.)

Example: 90S
             $-$50M
                40S

Write your Parity Score for Sex and Affection below.

Parity Score for Sex and Affection:
__________ ("S" or "M")

Step 11. The more similar your and your mate's Sex and Affection Scores, the smaller will be your Parity Score. Perfect parity has a score of zero. The range of possible parity scores is from 0 to 140S or 0 to 140M. Remember, if the Parity Score ends with an "S," this means you have greater power with the Sex and Affection Factor than your mate. If it ends with an "M," your mate has greater power. The chart below will be helpful in interpreting your Parity Score.

Parity Score Guide

"S" (Self) Scores

Zero		Perfect parity.
10S to	40S	Excellent balance in power.
50S to	80S	Medium balance, but you bring greater sex and affection power to the relationship than your mate.
90S to	140S	Dangerously unbalanced, unless your relationship includes compensating fair-trade assets. You have much greater sex and affection power than your mate.

"M" (Mate) Scores

Zero		Perfect parity.
10M to	40M	Excellent balance in power.
50M to	80M	Medium balance, but your mate brings greater sex and affection power to the relationship than you.

90M to 140M   Dangerously unbalanced,
               unless your relationship
               includes compensating
               fair-trade assets. Your mate
               has much greater sex and
               affection power than you.

(Now turn to the text of Chapter 9 on next page.)

Sex and Affection in Mating

How much lovemaking is correct? Which are the right procedures? How much kissing and hugging is "natural and desirable?" How many "I love you's" are required in a week?

When answering these questions it is important to keep two facts in mind:

(1) Sexual and affectionate satisfaction are completely personal issues, and
(2) Misconceptions about sex and affection abound.

Only you know how much sex and affection feels "right" and "satisfying" to you. You don't have to keep up with the Joneses or slow down to the pace of the Smiths.

People are quite different from one another, and there is no universal set of standards to measure your physical or emotional need satisfaction. In other words, no one is keeping count.

At the same time, however, misconceptions can distort our ideas about what is good for us and how we should go about achieving that which is desirable. For example, after many years of providing marriage therapy, we have heard an incredible variety of self-imposed sexual rules and regulations. These "do's" and "don't's" can be extremely and needlessly handicapping. Here are a few examples:

> "Lovemaking should be a hands-off experience. The natural way is not to use hands and mouth. Only the sex organs are to be used. The rest of it is disgusting. If nature had intended for us to use our hands, we'd have our sex organs on our fingers."
>
> "Women have to tolerate sex or their men will start getting it someplace else."
>
> "Of *course* I don't tell him what I want! He's

supposed to know. Where would be the romance in lovemaking if I had to give instructions? It's supposed to just 'happen.' It sure hasn't been though."

"Anyone who touches *himself* when having intercourse would have to be some kind of pervert . . . or trying to insult his partner."

"In my family we didn't say a lot of mushy things like 'I love you' and so forth. It's not my style. I don't need tenderness and I don't give it either."

As we said, the "right" amount of sex and affection for *you* is your decision. It's your body and your needs. However, be sure to think about and review your beliefs. See if you can cull out those that seem to be based on misconceptions and are not founded on fact.

Your Sex/Affection Score

Even though your own sensuality needs and interests are most important in deciding what is right for *you*, those who have higher scores have an advantage in society and in the mating marketplace. This doesn't mean that sexual athletes are welcomed everywhere. It has less to do with the sexual component and more to do with the ability to give others affection and reassurance.

Put it this way: Lovers are valued in society for their ability to give love. Sexual stamina and enthusiasm are of secondary importance and limited to the private domain.

Whether your score is High, Medium or Low, please give some thought to the emotional needs of others. Most people need more affection than we generally realize. We have the power to help others every day by showing them that we appreciate them and have good feelings for them.

You know how children respond to signs of affection.

Adults do too, although they are less likely to show it outwardly. We all need approval. One word of caution—you run the risk of becoming extremely popular and attracting many potential mates when people discover that you are a dispenser of love and affection. Don't say we didn't warn you.

Sex and Affection Parity

It is neither "right" nor "wrong" to enjoy sex once a month or fifteen times a week. It's up to you. However, it *would* be a major problem if you wanted the fifteen times a week and your mate had the other number in mind.

As with the other mating factors, it is vital that you have parity in sex and affection with your mate. Research clearly shows that similarity in sex drive, for example, results in greater compatibility.[1,2] The same is true for similarity in affectional behavior.

Whatever your sex/affection scores, you should look for a mate with similar scores. If you are unmated, compare your score with the "least acceptable" rating

---

Shhh . . . Daddy's Home

In Rome, around the time of Christ, the wife was considered a perpetual minor, with no ownership opportunities. If in the husband's eyes she did something wrong, he served as judge and jury and carried out the punishment. He could kill or sell into slavery any member of his family. Obviously, the family tip-toed when Daddy was home.

and think about whether any discrepancy might cause problems for you.

It is possible to work toward sex/affection parity in a marriage even when the partners have quite different interests. The secret is to communicate your needs and interests and listen carefully to what the other is saying.

If you like to be hugged, it is your responsibility to tell your partner how much and when.

Contrary to common opinion, the pleasure you derive from your physical relationship is *your own responsibility.* Only you can feel what is happening to your body, and it is up to you to communicate your desires to your mate. It doesn't work if you play the game of "Close your eyes and guess what I like." If you want it, whatever it may be, you must be willing to do a little teaching too.

If your mate is unwilling or somehow unable to give you the affection or sexual satisfaction you want, you have another important decision to make: Is this factor critical to your relationship? Are there other net worth assets that make the physical side less important by comparison?

Keep in mind that the physical side of a relationship often takes more time to "fine tune" than we realize. Some confusion and awkwardness about sex and affection should be expected in the first part of every mating relationship. The better the communication, however, the less time will be spent in the adjustment stage.

Incidentally, only rarely does the marriage therapist hear complaints about "too much affection" or "too much sex." Less than two percent of the women responding to a massive national survey indicated an interest in less sex.[3]

In conclusion, attitudes about sex and affection have been changing dramatically in the last fifteen or twenty years. The importance of showing love and affection is becoming more widely recognized as "not just for kids."

Women are less likely to accept a merely tolerant and passive role in sex, and both men and women are more aware of their capacity for sexual pleasure and satisfaction.

Marriages prosper when the partners have similar needs and interests concerning sexual frequency, types of sexual activities, and levels of affection. From the point of view of marriage success, there is only one standard for judging the quality of the sexual and affectional relationship between mates: their mutual satisfaction.

10

The Social Side of Mating

"IT'S NOT *what* you know, it's *who* you know."

Well, actually *both* are important to your future.

The *what* you know, your knowledge and skills, will determine much of your life, from career success to the kind of mate you will attract.

The *who* you know is extremely influential too, because it represents your entire social environment. It's more than the "social contacts" who can help you get better jobs or into special programs. Your social environment includes your parents, friends, business associates, relatives and, most importantly, your mate.

The people who surround you will influence just about every aspect of your life, including your self-concept, lifestyle, education, and what you do for fun.

Think about this for a minute: Haven't certain others always played a role in your important decisions? Usually parents, trusted friends, and other significant people help to shape our lives . . . even when we deliberately do the opposite of what they might want.

It is to your advantage to use the wisdom and experience of others, of course.

At the same time, since your social world is going to influence you so profoundly, why not select your social environment carefully?

Of course, you cannot choose your parents or family members, but you can select friends and career associates, and above all, your mate. And if you are already married, you need to give much thought to (a) the ways the two of you mutually influence each other and, (b) how other people influence both of you.

We'll return to this discussion after you have completed the tests dealing with the social factor in mating.

Note Carefully Whether the Tests and Instructions That Follow are for Unmated or Mated.

SOCIAL STATUS INVENTORIES

Test 1

MY SOCIAL STATUS
Unmated and Mated

Directions: This test is to be completed by everyone, both mated and unmated. Rate each of the following statements from 1 to 10 as it pertains to *you*. A rating of 1 means the statement is not at all descriptive of you. A 10 rating means it is a highly accurate statement about you.

1. _______ My career gives me high status.
2. _______ My friends generally possess high social status.
3. _______ People like to associate with me, as it reflects well on them.
4. _______ I live in a high status community.
5. _______ I have many friends.
6. _______ I am held in high esteem in the community.
7. _______ The way I live has a lot of "class."
8. _______ I am invited to belong to many organizations.
9. _______ Most of my friends are leaders.
10. _______ I am recognized as being well educated.
11. _______ I come from a family with high social status.
12. _______ In most groups I am a leader.
13. _______ I am considered to possess excellent social manners.

14. ______ I am often invited to parties and social occasions.
15. ______ In the future I'll have even higher social status.

(If you are unmated, proceed to Test 2. Mated go directly to Test 3.)

Test 2

THE SOCIAL FILTER
Unmated Only

Directions: The purpose of this test is to determine the level at which you filter out potential mates who fail to meet your minimum social standards. Which is the *lowest* level acceptable to you in a mate?

1. ______ Low — The mate I select can have low social status if other fair-trade assets are sufficient.

2. ______ Medium — I would not select a mate, irrespective of other assets, with less than average social standing.

3. ______ High — Notwithstanding any other assets a potential mate might have, I will only accept a mate with high social status.

(Unmated go directly to Test 3.)

Test 3

IMPORTANCE OF SOCIAL STATUS
Unmated and Mated

Directions: Both Unmated and Mated answer Question
A. Only Mated respond to Question B.

A. How much importance do you attach to social
status, as reflected by the motivation to
improve your social status level? (Check the
appropriate level.)

Low _______ Medium _______ High _______

(Levels are described below.)

B. How much importance does *your mate* seem to
attach to social status, as reflected by the
motivation to improve social status levels?
(Check the appropriate level.)

Low _______ Medium _______ High _______

(Levels are described below.)

Low: Social considerations are not important,
and it's not worth making an effort to
improve them.

Medium: Social status is important, and it's worth
making a reasonable effort to improve
it.

High: Social factors are of great importance,
and a major effort to achieve a high
level of social effectiveness is clearly
worthwhile.

*(Unmated proceed to Scoring Instructions on page 214. Mated
go to Test 4.)*

Test 4

MATE'S SOCIAL STATUS
Mated Only

Directions: This test examines your views of your *mate's* social status. Respond to each item as it relates specifically to your mate. Rate each of the following statements from 1 to 10. A rating of 1 means the statement is not at all descriptive of him/her. A 10 rating means it is a highly accurate statement concerning your mate.

1. _______ My mate's career provides high status.
2. _______ My mate's friends generally possess high social status.
3. _______ People like to associate with my mate, as it reflects well on them.
4. _______ My mate has many friends.
5. _______ My mate lives in a high status community.
6. _______ My mate is held in high esteem in the community.
7. _______ The way my mate lives has a lot of "class."
8. _______ My mate is invited to belong to many organizations.
9. _______ Most of my mate's friends are leaders.
10. _______ My mate is recognized as being well educated.
11. _______ My mate comes from a family with high social status.
12. _______ In most groups my mate is a leader.
13. _______ My mate is considered to possess excellent social manners.
14. _______ My mate is often invited to parties and social occasions.

15. _______ In the future, my mate will have even
            higher social status.
    _______ TOTAL SCORE

(Mated go directly to Scoring Instructions on page 216.)

Scoring Instructions

SOCIAL STATUS INVENTORIES
Unmated Only

Directions:

Step 1. Add the numbers in the Score Column in
        Test 1 (Social Status) on page 210.

        Place the Total Score here _________ .

Step 2. Using the conversion chart below, convert
        your Total Score from Step 1 to Scaled Points
        and place a (√) by the appropriate level:

Conversion Chart

Score from Step 1		*My Scaled Points*	*(√)*
15 to 41	=	10 Points	__________
42 to 68	=	20 Points	__________
69 to 95	=	30 Points	__________
96 to 122	=	40 Points	__________
123 to 150	=	50 Points	__________

Step 3. Look back at Test 3A (Importance of Social
        Status) on page 212.

        What was your rating? ________________
                              (Low, Medium, or High)

Convert this rating above to a number by using the following scale, and circle the appropriate number below:

Low = 1   Medium = 2   High = 3

Step 4. Multiply your Scaled Points (from Step 2 above) by 1, 2, or 3, depending on which number you circled in Step 3.

For example, if you have 30 Scaled Points in Step 2 and your level of importance score from Step 3 is Medium (2), then multiply 30 × 2 = 60.

Do your computations and then write your Social Status Score here:

Social Status Score _________ .

Step 5. In the conversion chart below, you will see how your Social Status Score rates in the mating marketplace. Place a (✓) by the appropriate level.

(✓)   *Social Status Level*

_________   10 to 50   =   Low Social Status
_________   60 to 100   =   Medium Social Status
_________   110 to 150 =   High Social Status

(Now turn to the text of Chapter 10, starting on page 221.)

Scoring Instructions

SOCIAL STATUS INVENTORIES
Mated Only

Directions:

Step 1. Add the numbers in the Score Column in Test 1 (Social Status) on page 210.

Place the Total Score here: ________

Step 2. Using the conversion chart below, convert your Total Score from Step 1 to Scaled Points and place a (√) by the appropriate level:

Conversion Chart

Score from Step 1		My Scaled Points	(√)
15 to 41	=	10 Points	________
42 to 68	=	20 Points	________
69 to 95	=	30 Points	________
96 to 122	=	40 Points	________
123 to 150	=	50 Points	________

Step 3. Add the numbers in the Score Column in Test 4 (*Mate's* Social Status) on page 214.

Place the Total Score here: ________

Step 4. Using the conversion chart below, convert your *Mate's* Total Score from Step 3 to Scaled Points and place a (√) by the appropriate level:

Conversion Chart

Score from Step 3		*Mate's Scaled Points*	(√)
15 to 41	=	10 Points	______
42 to 68	=	20 Points	______
69 to 95	=	30 Points	______
96 to 122	=	40 Points	______
123 to 150	=	50 Points	______

Step 5. Look back to Test 3A (Importance of Social Status) on page 212.

What was your rating? ______________
(Low, Medium, or High)

Convert this rating to a number by using the following scale, and circle the appropriate number below:

Low = 1   Medium = 2   High = 3

Step 6. Return to Test 3B on page 212. What was your *Mate's* rating?

Mate's rating: ______________
(Low, Medium, or High)

Convert this rating to a number by using the following scale, and circle the appropriate number below:

Low = 1   Medium = 2   High = 3

Step 7. Next we will compute Your Social Status Score. Multiply *My* Scaled Points (Step 2) by 1, 2, or 3, depending on which number you circled in Step 5.

For example, if you have 30 Scaled Points from Step 2 and you have circled 2 (Medium) in Step 5, then multiply 30 × 2 = 60S. We have placed an "S" next to the

score. It stands for "Self," as this is your own Social Status Score.

Do your computations and write your Social Status Score here:

My Social Status Score _________ s (Self)

Step 8. Now we need to compute your *Mate's* Social Status Score. Multiply your *Mate's* Scaled Points (from Step 4) by 1, 2, or 3, depending on the Importance Rating circled in Step 6.

For example, if your mate has 20 Scaled Points (in Step 4) and the Level of Importance Score (from Step 6) is 2 (Medium), then multiply $20 \times 2 = 40M$. We have included an "M" next to the score. It stands for "Mate," as this is your *Mate's* Social Status Score.

Now do your computations and write your *Mate's* Social Status Score here:

Mate's Social Status Score _________ M (Mate)

Step 9. The conversion chart below translates the Social Status Score for you and your mate into levels for easy comparison.

Place an "S" (Self) by the appropriate level for *your* score from Step 7. Place an "*M*" (Mate) by the appropriate level for your *Mate's* score from Step 8.

"S" or "M"   Social Status Level

_________   10 to 50   Low Social Status
_________   60 to 100   Medium Social Status
_________   110 to 150   High Social Status

Step 10. Look at your Social Status Score in Step 7 and your *Mate's* Social Status Score in Step 8. Which is larger? To find your Parity Score for Social Status, subtract the smaller number from the larger number and carry to your answer the "S" or "M" from the larger number.

For example, if your Social Status Score is 90S and your Mate's Social Status Score is 50M, your Parity Score for Social Status is 40S. (See example below.)

Example:

$$\begin{array}{r} 90\text{S} \\ -50\text{M} \\ \hline 40\text{S} \end{array}$$

Write your Parity Score for Social Status below:

Parity Score for Social Status __________ ("S" or "M")

Step 11. The more similar your and your mate's Social Status Scores, the smaller will be your Parity Score. Perfect parity has a score of zero. The range of possible parity scores is from 0 to 140S or 0 to 140M. Remember, if the Parity Score ends with an "S," this means you have greater power with the Social Status Factor than your mate. If it ends with an "M," your mate has greater power. The chart below will be helpful in interpreting your Parity Score.

Parity Score Guide

"S" (Self) Scores

Zero	Perfect parity.
10S to 40S	Excellent balance in power.

50S to 80S   Medium balance, but you
bring greater social power
to the relationship than
your mate.

90S to 140S   Dangerously unbalanced,
unless your relationship
includes compensating
fair-trade assets. You have
much greater social power
than your mate.

"M" (Mate) Scores

Zero   Perfect parity.
10M to 40M   Excellent balance in power.
50M to 80M   Medium balance, but your
mate brings greater social
power to the relationship
than you.

90M to 140M   Dangerously unbalanced,
unless your relationship
includes compensating
fair-trade assets. Your
mate has much greater
social power than you.

(Now turn to the text of Chapter 9, starting on the next page.)

Social Factors in Mating

Remember the social cliques in high school? Do you remember who went out with whom, which kids were considered "socially acceptable," by various subgroups, and how the various social levels seemed to emerge? Did your school have "greasers," "jocks," "brains" (sometimes called "fags"), "hippies" (or "heads"), and so forth?

Society after high school has a social class system too, although it's not quite polite to talk about it. Most of the levels are based on what a person does for a living, and an endless variety of smaller groups is formed around special interests at the various levels. For example, people interested in automobile restoration will rejuvenate Ferraris at one level and Firebirds at another.

The good news is that with only a few exceptions, the various social levels and groups are relatively open. That is, it is possible to move from one social level to another. It requires some planning and effort, however, and the question for each individual is whether it is worth the time and energy.

The higher social levels have all sorts of advantages, of course. If you will think back to our analogy of the social stairway, you will remember that people who have higher net worths can and will trade and share those assets. If you'd like to operate at that level, it requires that your own net worth be high enough.

Unmated people need to decide whether they should select a mate with parity and then either stay at that level or improve together, or postpone mating until they have achieved enough additional assets to deserve a mate from a higher level. For instance, we all have seen countless examples of romances which broke up after a time because one person went to college or added new assets while the other marked time. Young women have been especially vulnerable to losing a lover when early parity was lost after the male became more successful.

If you decide to move to higher social levels, keep these factors in mind:

(1) Moving to a higher level should be based on your improved net worth.

(2) You should genuinely share the values and interests of the people at the different social levels.

(3) You should select the social environment carefully, recognizing that there literally are millions of groups from which to choose.

Let's consider these points a bit further. When you improve your social status by increasing your net worth (and probably developing a speciality or two), you will have earned the respect of equally effective people higher on the social stairway (discussed in Chapter 3). They will welcome you and want to trade with you.

Your new status will, however, cost you some of your previous associates. Once you have achieved a higher net worth, parity is lost and they will begin to be uncomfortable with you . . . and vice versa.

Many of the people you used to "hang out" with, chatting amiably and sharing little gossip, will become, to your dismay, quite boring. Suddenly you and they will have relatively little in common and you'll find yourself yawning when you are around them. The reason? You will have outgrown them as surely as if they were still in the seventh grade.

Imagine how you and your mate would feel if one of you "outgrew" the other! This has happened to millions of marriages, with predictable consequences. Parity must be maintained if you want your relationship to succeed. If you plan for mutual growth and continued parity, there is less danger that you will find yourself on one level and your mate on another. The key is to move up the social stairway together. You do not necessarily have to share the same social groups or even the same friends, although it can be extremely beneficial to do so. Just be sure that your parity is protected.

Puffery

Moving a person into higher levels of social effectiveness is almost always worth the effort. But have you seen people struggling to be impressive in order to be "accepted" by the "right" people? They memorize facts from the latest *Newsweek* or *Time* to drop during dinner conversation, pretend to love ballet and lobster bisque because they think it's classy, or struggle to pronounce words with a more desirable accent.

But the use of what we call "puffery" is not an effective strategy. It requires a great deal of effort to keep up a phony front, feigning interests and opinions which are not really believed. If you wish to move to more powerful social circles, you should use strategies which work. And effective strategies are based on increasing your net worth.

In the same way, your parity with your mate or prospective mate must be based on your real mating power. You can and should include your plans and motivation in your net worth estimates, and you should be willing to discuss those plans with your mate. Again, try to represent yourself accurately and encourage your mate to do the same. The importance of social parity is too great to play pretending games with each other.

The Social Power of Your Mate

As you make your decisions about your goals in life, keep in mind how influential your mate will be in your efforts to achieve them. Whatever you want to do, you should select your mate with this awareness. If he or she does not share your dreams and aspirations, you are either not likely to achieve them or the marriage will not succeed.

When you select a mate, in a very real sense you are choosing a small but extremely powerful social environment. Want to be financially successful? Marry someone who will share your motivation? Have dreams of seeing the world? With the right mate, it's going to happen. Do you especially want to live with joyful exuberance? Crave adventures aplenty? Prefer the philosophical contemplative life? Then select your social environment, and especially your mate, with the greatest of care.

Social scientists of all kinds, and barbers and yentas too, agree that we live in a social world which by nature, is going to modify us and influence our destinies.

In other words, we are not only *known* by the company we keep, we are a *product* of that company!

It's not that we are powerless pawns of the people around us, however. We have the capacity to influence them also. Just as importantly, we can *choose* the groups we want to be around! In other words, as long as we are going to be influenced, we should decide which directions we want to go and become involved with the ones who will help us along.

Most of us have a wonderful capacity to rise to any occasion, but the occasion needs to make demands of us if we are to progress. To spend our time in an undemanding environment is to reduce the rate of growth to a feeble shadow of what it might be. By choosing a mate who will help establish a dynamic environment, we have a greater chance of making the environment work to our advantage.

Social Status and Your Specialty

Your social level, with its advantages and disadvantages, is determined by your net worth. As we indicated earlier, one excellent strategy for increasing net worth and social level is to develop a specialty which is highly valued.

It can be just about any specialization, including, for example, such areas as antiques restoration, musical talent, an unusual business or travel experience, the ability to amuse, athletic prowess, or special knowledge about anything.

What specialty should *you* develop? If you have any interest or talents already, they more than likely could be developed. You should become an expert in something you enjoy, and be certain to choose areas in which you have an opportunity to excel. The more specialized and unusual the area, the more chance you have to develop a unique talent or body of knowledge, and therefore be admitted to higher social status. The more desirable your specialty, the further it will take you.

Social Skills

Your specialty should be supplemented by a few essential social skills to steady your climb up the social stairway.

Appearing confident is a must. There can be no toe scraping modesty, with its shy little "Oh, it's not much. Just a silly little idea I worked on in my spare time. It needs a lot of work yet."

You will need to speak with enthusiasm about yourself and what you have to offer. If you don't believe in yourself, how can you expect others to? We might add, however, that when you have earned a higher net worth and have a specialty others value, you naturally will feel much more enthusiastic about your own worth.

Another needed social skill is the ability and willingness to trade assets and, especially, information. You have something that others value, and you must let them know that you are in the marketplace and willing to share. Misers have little chance of making significant net worth gains.

We'll count attractiveness as an essential social asset too. Whether you were blessed with natural beauty or not, appearance is important. As we indicated in Chapter 7, it particularly involves attention to grooming and the clothes you wear. Successful people do "judge a book by its cover" because they recognize that looking after your appearance is one of the most effective ways of letting others know that you are important enough to be taken seriously. It also shows you are willing to make the extra effort to manage your own destiny.

Finally, associate with other socially successful people to observe their effective interpersonal skills. You and your mate can work together to develop new and better social skills, and any effort you make together will be productive, not only for each of you, but for your relationship as well.

11

What Do You Value?

WHAT IS IMPORTANT to you in life?

What do you value? Do you thrive on variety and change, or do you prefer a more steady and predictable lifestyle? Is independence high on your list of priorities, or companionship, or raising a family?

Your answers to these and other questions concerning your essential values will determine the directions your life should take. Your values also should influence you in a major way as you make your mating decisions.

After all, if what you value is not a major part of your life, what satisfaction is possible? And if your mate doesn't share your values, you are far less likely to live happily and productively.

In this chapter, we provide two tests. Both are extremely important to successful mating and concern the values you and your mate bring to your marriage relationship. These tests are quite different from others you've taken in MATING, so please follow the instructions carefully.

Note Carefully Whether the Tests and Instructions That Follow are for Unmated or Mated.

MARRIAGE VALUES INVENTORIES

Test 1

MARRIAGE READINESS
Unmated and Mated

Directions: A vital determinant of marriage happiness and success is the importance attached to married life. Respond as honestly as you can to the following items. Indicate which of the two choices is more important to *you*. Place a (✓) in Column X next to Option A or B for each of the paired statements.

If you have a mate, also place a (✓) in Column Y next to Option A or B for each item. Make your decision based on which option you believe is more important to your *mate*.

Column X *My* *Choice*	Which statement more nearly reflects your personal opinions and priorities? Which reflects your mate's?	*Column Y* *My Mate's* *Choice*
1. ___A	I wouldn't feel alive unless I married.	___A
___B	Give me liberty or give me death.	___B
2. ___A	My "favorite things to do" are typical of married people.	___A
___B	My hobbies and interests are those of a single person.	___B

3. ___A   Being married would give me more of the security I want.   ___A

    ___B   Being single would give me more of the career opportunities I want.   ___B

4. ___A   I'd love to vacation with my spouse.   ___A

    ___B   I'd love to vacation with different people at different times.   ___B

5. ___A   Married, therefore sexually content.   ___A

    ___B   Single, therefore sexually content.   ___B

6. ___A   I prefer the security of married life.   ___A

    ___B   I prefer the adventure of single life.   ___B

7. ___A   I want sex and affection from one reliable person.   ___A

    ___B   I want a variety of lovers.   ___B

8. ___A   Marriage is underrated.   ___A

    ___B   Marriage is overrated.   ___B

9. ___A   I'm willing to work to make my spouse happy.   ___A

    ___B   I don't want to be responsible for my mate's happiness.   ___B

10. ___A   It is morally more correct to live a married life.   ___A

    ___B   I'd like to do things when I want, without family constraints.   ___B

11. ____A   I'm ready to share my life and my   ____A
              credit cards.
    ____B   I'm ready to share good times, but   ____B
              let's not get too serious.

12. ____A   I believe that people should be   ____A
              married.
    ____B   I think people are happier single.   ____B

13. ____A   Sex is best with one's husband or   ____A
              wife.
    ____B   Sex is too delicious to limit it to   ____B
              one person.

14. ____A   My career plans can tolerate or   ____A
              even benefit from being married.
    ____B   My career plans would benefit   ____B
              from being single.

15. ____A   I prefer for most of my friends to   ____A
              be married.
    ____B   I prefer for most of my friends to   ____B
              be single.

16. ____A   A marriage can be a beautiful   ____A
              experience.
    ____B   A marriage can be a trap.   ____B

17. ____A   I'd like to be able to know who I'll   ____A
              be sleeping with for the rest of my
              life.
    ____B   I don't want to know who I'll be   ____B
              sleeping with for the rest of my
              life.

18. ___A___   Today's smart people are getting married.   ___A___

    ___B___   Today's smart people are single and enjoying it.   ___B___

19. ___A___   The pleasures of marriage are wonderful.   ___A___

    ___B___   The risk of a bad marriage more than offsets any pleasure.   ___B___

20. ___A___   "Love" and "marriage" go together.   ___A___

    ___B___   "Love" and "independence" go together.   ___B___

21. ___A___   I don't mind sharing bank accounts and expenses.   ___A___

    ___B___   I'm too independent to enjoy sharing my money with a spouse.   ___B___

22. ___A___   I want to pour all my love into a permanent relationship.   ___A___

    ___B___   I'm not happy with an exclusive love relationship.   ___B___

23. ___A___   The economic advantages of being married are important to me.   ___A___

    ___B___   I like the idea of being single so I can spend my money as I wish.   ___B___

24. ___A___   To me, being married is important from a moral/religious standpoint.   ___A___

    ___B___   To me, morality and religion have little to do with marital status.   ___B___

25. ____ A   Lovemaking with one's spouse is   ____ A
                 certainly best.
    ____ B   The many sexual opportunities of   ____ B
                 being single are a great advantage.

26. ____ A   My greatest chance for personal   ____ A
                 growth is through marriage.
    ____ B   My greatest chance for personal   ____ B
                 growth is through independence.

27. ____ A   I'd prefer to share the risks of life   ____ A
                 with a spouse.
    ____ B   My privacy is too important for   ____ B
                 me to enjoy being married.

28. ____ A   I want a reliable lifetime   ____ A
                 relationship in marriage.
    ____ B   I'd rather live free and easy. . .and   ____ B
                 singly.

29. ____ A   I'd be willing to make major   ____ A
                 changes, such as moving to
                 another state, to help a spouse.
    ____ B   I'm not ready to move, or make   ____ B
                 other major changes, to help a
                 spouse.

30. ____ A   Marriages are made in heaven.   ____ A
    ____ B   Marriages are too often made in   ____ B
                 hell.

31. ____ A   My recreational interests are   ____ A
                 family-oriented.
    ____ B   My recreational interests are for   ____ B
                 singles.

32. ____**A**   I'd enjoy the security involved in being married.   ____**A**

    ____**B**   I'd prefer the freedom of being single.   ____**B**

33. ____**A**   I would love to have children.   ____**A**

    ____**B**   I don't want the responsibility of a family.   ____**B**

34. ____**A**   Being married offers joint tax returns and other financial benefits.   ____**A**

    ____**B**   I prefer to spend my money the way I want to.   ____**B**

35. ____**A**   I prefer sex and affection with my spouse for life.   ____**A**

    ____**B**   I prefer love affairs with many people.   ____**B**

36. ____**A**   I prefer the steady companionship of being married.   ____**A**

    ____**B**   I prefer the excitement of companionship with many lovers.   ____**B**

37. ____**A**   I'd enjoy spending leisure time with my spouse.   ____**A**

    ____**B**   I like spending my leisure time as I please.   ____**B**

38. ____**A**   If I have to choose between career advancement and marriage, I choose marriage.   ____**A**

    ____**B**   If I have to choose between career advancement and marriage, I'll take career, thank you.   ____**B**

39.   **A**   Married people can count on   **A**
                 sexual satisfaction from a trusted
                 partner.
  B   Single people can find lovers from   **B**
                 an endless pool of possibilities.

40.   **A**   Marriage requires major   **A**
                 compromises and I'm willing to
                 make them.
  B   Marriage requires major   **B**
                 compromises and I'm not yet
                 ready to make them.

(Unmated and Mated proceed to Test 2.)

Test 2

VALUES HIERARCHY
Unmated and Mated

Directions: The values you bring to a marriage are of major importance to its success. Respond to each of the items as honestly as you can.

The central question to be answered in this test is this: Which of the two options is of greater importance in a marriage? In Column X indicate with a (√) which of the two statements comes closer to your own value. *Unmated should complete Column X only.* If you evaluate your spouse or a potential mate, use Column Y to indicate with a (√) which of the two paired options you believe is closer to your mate's values.

It is important to complete all items in a column, even though a few choices may be very difficult for you to make. Making decisions about what we value is never easy, but it's worth the effort.

Column X *My* *Values*	Which is more important in a marriage?	*Column Y* *Mate's* *Values*
1. G____	Economic advantages of being married.	____ G
	or	
I____	Sharing of affection with one another.	____ I
2. C____	Feelings of security and safety in marriage.	____ C
	or	
J____	Social opportunities of married life.	____ J
3. E____	Sharing of household and daily living tasks.	____ E
	or	
F____	Moral and religious aspects of marriage.	____ F
4. B____	Reliable companionship within marriage.	____ B
	or	
J____	Social opportunities of married life.	____ J
5. G____	Economic advantages of being married.	____ G

or

    H _____   Respect and status from being   ____ H
married.

6.   A _____   Satisfaction in having family,   ____ A
raising kids.

or

    J _____   Social opportunities of married   ____ J
life.

7.   F _____   Moral and religious aspects of   ____ F
marriage.

or

    G _____   Economic advantages of being   ____ G
married.

8.   C _____   Feelings of security and safety in   ____ C
marriage.

or

    H _____   Respect and status from being   ____ H
married.

9.   D _____   Sexual satisfaction in married   ____ D
life.

or

    E _____   Sharing of household and daily   ____ E
living tasks.

10.   I _____   Sharing of affection with one   ____ I
another.

or

    J _____   Social opportunities of married   ____ J
life.

11.   A _____   Satisfaction in having family,   ____ A
raising kids.

or

  I ____   Sharing of affection with one   ____ I
        another.

12.   B ____   Reliable companionship within   ____ B
        marriage.

or

  I ____   Sharing of affection with one   ____ I
        another.

13.   C ____   Feelings of security and safety in   ____ C
        marriage.

or

  G ____   Economic advantages of   ____ G
        being married.

14.   D ____   Sexual satisfaction in married   ____ D
        life.

or

  F ____   Moral and religious aspects of   ____ F
        marriage.

15.   E ____   Sharing of household and daily   ____ E
        living tasks.

or

  H ____   Respect and status from being   ____ H
        married.

16.   B ____   Reliable companionship within   ____ B
        marriage.

or

  G ____   Economic advantages of being   ____ G
        married.

17.   A ____   Satisfaction in having family,   ____ A
        raising kids.

or

H _____   Respect and status from being   ____ H
          married.

18.   E _____   Sharing of household and daily   ____ E
          living tasks.

or

      J ___   Social opportunities of married   ___ J
          life.

19.   C _____   Feelings of security and safety in   ____ C
          marriage.

or

      F ___   Moral and religious aspects of   ___ F
          marriage.

20.   D ___   Sexual satisfaction in married   ___ D
          life.

or

      I ___   Sharing of affection with one   ___ I
          another.

21.   A ___   Satisfaction in having family,   ___ A
          raising kids.

or

      B ___   Reliable companionship within   ___ B
          marriage.

22.   H ___   Respect and status from being   ___ H
          married.

or

      I ___   Sharing of affection with one   ___ I
          another.

23.   C ___   Feelings of security and safety in   ___ C
          marriage.

or

D _____   Sexual satisfaction in married   _____ D
         life.

24. A _____   Satisfaction in having family,   _____ A
              raising kids.

or

     F _____   Moral and religious aspects of   _____ F
              marriage.

25. C _____   Feelings of security and safety in _____ C
              marriage.

or

     E _____   Sharing of household and daily   _____ E
              living tasks.

26. A _____   Satisfaction in having family,   _____ A
              raising kids.

or

     D _____   Sexual satisfaction in married   _____ D
              life.

27. B _____   Reliable companionship within   _____ B
              marriage.

or

     H _____   Respect and status from being   _____ H
              married.

28. B _____   Reliable companionship within   _____ B
              marriage.

or

     D _____   Sexual satisfaction in married   _____ D
              life.

29. F _____   Moral and religious aspects of   _____ F
              marriage.

or

     J _____   Social opportunities of married   _____ J
              life.

30.   E___   Sharing of household and daily ___ E
             living tasks.
                          or
      G___   Economic advantages of being   ___ G
             married.

31.   A___   Satisfaction in having family, ___ A
             raising kids.
                          or
      E___   Sharing of household and daily ___ E
             living tasks.

32.   B___   Reliable companionship within ___ B
             marriage.
                          or
      F___   Moral and religious aspects of ___ F
             marriage.

33.   C___   Feelings of security and safety in ___ C
             marriage.
                          or
      I___   Sharing of affection with one ___ I
             another.

34.   D___   Sexual satisfaction in married ___ D
             life.
                          or
      H___   Respect and status from being ___ H
             married.

35.   G___   Economic advantages of being   ___ G
             married.
                          or
      J___   Social opportunities of married ___ J
             life.

36. A____   Satisfaction in having family,   ____ A
raising kids.

or

   C____   Feelings of security and safety in   ____ C
marriage.

37. B____   Reliable companionship within   ____ B
marriage.

or

   E____   Sharing of household and daily   ____ E
living tasks.

38. D____   Sexual satisfaction in married   ____ D
life.

or

   G____   Economic advantages of being   ____ G
married.

39. F____   Moral and religious aspects of   ____ F
marriage

or

   I____   Sharing of affection with one   ____ I
another.

40. H____   Respect and status from being   ____ H
married.

or

   J____   Social opportunities of married   ____ J
life.

41. A____   Satisfaction in having family,   ____ A
raising kids.

or

   G____   Economic advantages of being   ____ G
married.

42. B____   Reliable companionship within   ____ B
marriage.

or

C _____   Feelings of security and safety in ____ C
marriage.

43.   D____   Sexual satisfaction in married ____ D
life.

or

J____   Social opportunities of married ____ J
life.

44.   E____   Sharing of household and daily ____ E
living tasks.

or

I____   Sharing of affection with one ____ I
another.

45.   F____   Moral and religious aspects of ____ F
marriage.

or

H____   Respect and status from being ____ H
married.

(Unmated and Mated turn to the text of Chapter 11 on next page.)

The Value of Similar Values

"It was my first day in the office, and when I saw her desk, I stopped dead in my tracks. It had a picture of a Cessna 180, which is my favorite airplane, a picture of an MG TD, which was my first and most beloved car, a copy of the novel I was reading that week, and a calendar featuring my home state!

"I didn't know whose desk it was, I just knew we were going to become friends. Well, we certainly did. After we actually met, we fell completely in love. We had so much in common."

Researchers discovered many years ago that the people we like most are the ones who are most like us.[1,2] Our friends and lovers almost always have similar likes and dislikes and believe as we do. Their politics and social attitudes are likely to be much the same as ours. In most cases, they will have the same religious beliefs and when they don't, there is a high risk of divorce.[3]

Why are people with similar values attracted to each other? One reason is that they are comfortable with each other's ideas and values, because they are so much like their own.[4,5] "You don't have to be on guard," as one client put it. "You can really be yourself and know that you will be understood and appreciated for it."

Affirmation as a Strategy

A second reason is that similar people give each other affirmation and acceptance.[6,7] We all love to hear people say, "You're right," "I agree," "Your position is correct," and "Good thinking!" Most of all, we love to hear YES!

The person who gives all this affirmation very quickly becomes an extremely valuable ally, probably a friend, and possibly a lover and mate. The more one affirms another's worth, the more he will be appreciated. And if two people affirm the *other's* values and beliefs, they have an excellent basis for a superb relationship.

People who make a living getting others to agree with them are familiar with the principle of mutual affirmation. They range from executives who learned the strategy in prestigious management seminars to the gifted salespeople who seem to have been born with all the right techniques.

If you want to develop a relationship with someone in whom you have an interest, you should make a special effort to communicate the value you attach to various qualities. Which kinds of qualities? It's nice to tell people you like their tie or shoes, but to develop a solid relationship, let them know you approve of what they value.

Founding Fathers

True to the promise he had given his dying wife Martha, Thomas Jefferson never remarried. That didn't keep him from enjoying the company and comforts of one of his teenage slaves, Sally Hemings. Their five children, we hope, reaped the benefits of Jeffersonian democracy.

As a young firebrand, Patrick Henry stood up and declared, "Give me liberty or give me death." Evidentally he believed in liberty for his wife as well, for when he was given death, his will stipulated that she could hold his estate only so long as she was a free woman and did not remarry.

Gouvenor Morris was of a different mind. At his death his widow inherited half of his estate. The will provided that she could have the other half the moment she married again.

You will, in essence, be affirming what they believe and you, in turn, will be affirmed for doing it. Keep it up and the affirmation of values will become reciprocal, or two-way.

In sum, we feel affection and love for those who affirm what we believe and value. If you are attracted to someone with similar values, you need not wait passively for a relationship to develop. By speaking up and approving, you will start a relationship off on a strong and positive footing.

The important people in our lives are chosen, then, because they see the world as we do, have the same values we have, and they give us positive feedback about ourselves and our ideas.

What are Values?

Value similarity is just as important to a marriage as to a friendship or other close relationship. But what is a value? We have used the term in a broad sense rather than a narrow one. It includes our beliefs, attitudes, and interests. Values also are the principles we use to determine what is right or wrong, desirable or undesirable.

Marriages can get into trouble if the partners don't agree on most of what they value. Imagine what would happen in a marriage if one mate placed a higher value on honesty than success and the other subscribed to the "Winning is everything" approach.

One couple we interviewed had a terrific clash over values when she placed a premium on freedom and independence and he insisted on stability and security:

Sharon vs. James

Sharon already was, at the age of 26, a spectacular success in the corporate world. She was blessed with the direct and forceful gaze, confidence, and social skills so valued in the business community, and most important of all, she was intelligent and willing to work very late hours to see a job through.

James was successful also, with his law practice starting to blossom after lean times, and he and Sharon looked forward to children in perhaps six or seven years. They lived well, enjoying the "perks" that their efforts provided.

When Sharon announced the exciting news that she had been offered a fantastic opportunity 1500 miles away, and that they would need to move within the next two months, James was incredulous. "You're crazy!" he bellowed. "We're not going anywhere! My practice is here, my family and friends are here, and we can't give up what we have just so you can be a star."

She was equally agitated, not so much about possibly losing the chance at such a splendid promotion, but because she thrived on change, challenge, moving about the country, meeting new people, and taking giant steps from one corporation to another. Routine and security were not high on her list of values. And she was *very* disappointed with James.

What would you do? We all make value decisions every day, not all of the magnitude in the example above, but significant nevertheless. Your actions and plans, the sacrifices you make for others, the way you spend your leisure time—all of these and so much more are shaped and molded by values.

Have you ever been involved in the "what if" exercises used in values clarification? The ones where you hear "What if there were nine people and only room for five in a bomb shelter . . . ?" or "What if a vaccine were available for a dread disease, but only enough existed for half the people? Which would you save?"

Such exercises raise some exciting arguments and real questions about what we value and why.

It's all a question of priorities. We may value peace, but be willing to fight for freedom because we value freedom more than peace. We value leisure time, but work because we value the money more. We all have value ladders or hierarchies established for our lives, although many of them are only hazily formulated or are inconsistent.

Marriage Values

The values which are most important to this book have to do with mating rather than world peace or the meaning of life. From a practical standpoint, which do you value more: having children or building your career; the reliable companionship of marriage, or the variety of sexual partners more likely if you are single?

In this chapter we are concerned with two special issues which have to do with values:

(1) Whether you place a higher value on being married or on a singles lifestyle, and
(2) How your values compare with your present or potential mate.

Both will have far-reaching implications for you.

Marriage Readiness

"They just weren't ready for marriage," is a phrase heard so often after a divorce that one must suspect it means something important. Upon closer inspection, it becomes apparent that it refers to the value placed by one or both spouses on being married. And it certainly is important; it is potentially more lethal to a marriage than any of the other five mating factors.

Put another way, a marriage can survive even if the

couple has little money, low social status, poor personal effectiveness and the sexuality and attractiveness of two turnips. It cannot succeed if either spouse does not value being married.

There are single people who have high readiness for marriage, but there also are married people who do not place much value on being married. The Marriage Readiness Test can help you decide whether marriage is appropriate for you. If you are married and discover that you and/or your mate have "singles" values, then steps will need to be taken if you are to protect and improve your relationship. Professional consultation would very likely be beneficial.

Comparing Marriage Values

The more *similar* the marriage values of a husband and wife, the greater the marital satisfaction.[8] But unlike many of the other mating factors, the issue is not *amount* but relative *position* in an hierarchy. That is, no one can say that one set of values is worth more than another, but a relationship is strongest when the mates *agree* on the importance of a variety of values.

The Values Hierarchy Test is unique. It has been especially developed to accomplish two purposes: (1) It provides information concerning the importance you and your mate attach to selected values, and (2) It compares the values hierarchies of the marriage partners to determine how similar or different their values really are. If you are unmated, this test will provide you with your values hierarchy which will be useful when you later consider getting married and you want to do some serious comparing.

Now it's time to score your two tests and examine your marriage readiness score and values hierarchy.

Scoring Instructions

MARRIAGE READINESS
Unmated Only

Directions:

Step 1. Return to the Marriage Readiness Test 1 on page 228.

Step 2. Count all of the "A" responses and write the number of "A" responses here: _________ .

Step 3. Determine your "Score Level" by comparing the number of "A" responses from Step 2 above with the following scale. Then place a (√) next to your level.

(√)

0 – 20	=	Low Readiness	_______
21 – 30	=	Medium Readiness	_______
31 – 40	=	High Readiness	_______

Step 4. Now you are ready to review your Marriage Readiness Profile. Turn to the appropriate page listed below:

Low Readiness	Page 252
Medium Readiness	Page 253
High Readiness	Page 254

Scoring Instructions

MARRIAGE READINESS
Mated Only

Directions:

Step 1. Return to the Marriage Readiness Test 1 on page 228.

Step 2. Count all of the "A" responses in the "X" Column and write the number here: ______ .

Step 3. Determine your "Score Level" by comparing the number of "A" responses from Step 2 above with the following scale. Then place a (√) next to your level.

(√)

0 – 20	=	Low Readiness	______
21 – 30	=	Medium Readiness	______
31 – 40	=	High Readiness	______

Step 4. Now count all of the "A" responses in the "Y" Column and write the number here: ______ .

Step 5. Determine your *Mate's* "Score Level" by comparing the number of "A" responses from Step 4 above with the following scale. Then place a (√) next to your mate's level.

(√)

0 – 20	=	Low Readiness	______
21 – 30	=	Medium Readiness	______
31 – 40	=	High Readiness	______

Step 6. Now you are ready to review your Marriage Readiness Profile as well as the profile for your mate. Turn to the appropriate page(s) listed below:

Low Readiness   Page 255
Medium Readiness   Page 256
High Readiness   Page 257

MARRIAGE READINESS
Unmated

Low Marriage Readiness

Your values are heavily weighted toward a single life-style. The Marriage Readiness Principle is this: People should marry only when the value they attach to married life is greater than the value they attach to being single. In addition, of course, they should marry a parity mate with whom they can build a lifetime relationship.

Even if you have found a parity mate, the value you attach to married life is not enough to insure a successful marriage. Do not marry on the assumption that once married your values will change. They may or they may not. And you should not marry on the basis of other people's expectations or their views of the social desirability of being married. Your values are the ones that count when it comes to mating.

Be careful about falling into the trap of using your "less interested" status to make excessive demands on your prospective mates. It is, of course, unfair and destroys relationships before they have a chance to develop.

Marriage readiness can develop rapidly. You may grow tired of those things which characterize being single. On the other hand, you may come to appreciate more those things which are associated with being married. Continue to review your marriage readiness from time to time, but do NOT marry until your values will support it.

(Turn to Test 2 Scoring Instructions on page 258.)

Medium Marriage Readiness

Your values blend support for singles and married lifestyles. They carry with them a level of risk if you choose to marry at this time.

Probably the most important issue for the individual with medium marriage readiness is the readiness of a potential mate. A mate with high readiness can add a dimension of stability to the relationship which increases the likelihood of success. On the other hand if your potential mate had only a medium or low readiness too, the combination would dramatically reduce the already uncertain potential for such a marriage.

Be alert to the dangers of the less marriage-ready partner making excessive demands for concessions. That is, the demands of some mates say, in effect, "Do what I want or I'm getting out. I don't really want to be married as much as you do anyway." This is a destructive gambit, and should be recognized as such.

A second major issue for those with medium marriage readiness is the parity fit between them and their potential mate. This level of readiness provides little cushion against the costs of low parity. Look at it this way: High parity is likely to lead to a successful marriage even if one or both of the partners have only medium marriage readiness. As the parity levels decrease, the greater the need for high readiness.

(Turn to Test 2 Scoring Instructions on page 258.)

High Marriage Readiness

Your values strongly support married life. Your only challenge will be to find a parity mate. Keep in mind that the values readiness of a potential mate also will affect the success of any relationship. Be on guard against projecting your own high level of marriage readiness onto a potential mate. There is also a danger that you will let your readiness to marry influence you into entering into a non-parity relationship. The key for you is to realize that although you are ready, you still need to choose wisely.

(Turn to Test 2 Scoring Instructions on page 258.)

MARRIAGE READINESS
Mated

Low Marriage Readiness

Those who are married but have a low marriage readiness usually place too little value on marriage to be good partners. It is not that they want to be difficult. Rather, they do not value highly those things which marriage has to offer. When both partners have low readiness levels, the chances for the relationship are extremely dim, and if the marriage also suffers from nonparity, the odds are even longer.

The best hope for such marriages is honest communication between the spouses. The survival of their relationship is at stake. They must find a way to accommodate some of the non-marriage values of the partners without destroying the relationship.

There is reason for hope. Values change with time and learning. The key is for the couple to hold the marriage together, strive for parity, and jointly build the relationship as they learn to value it. Marriage therapy would be appropriate, of course, and making the time for much-needed discussion and planning is vital.

Both partners must be especially careful of the unfair advantage held by the partner with less interest in the marriage. In most arguments he or she could make inordinate demands or threaten to dissolve the relationship (with less to lose). This is an extremely destructive tactic and few marriages will survive it.

(Turn to Test 2 Scoring Instructions on page 259.)

Medium Marriage Readiness

Relationships characterized by medium marriage readiness are sustainable, particularly if *both* partners have at least a medium level of readiness. However, such relationships have little cushion against the problems and pressures of daily living. The danger to your marriage would be compounded if your partner has a low readiness for marriage or the relationship suffers from non-parity in net worths.

To the extent that your marriage meets with success on a day-to-day basis, you and your spouse are likely to place a higher value on it. We all value more highly those aspects of our lives in which we are successful. This will require discussion and planning together, but it's worth the effort. Both large and small successes and pleasures will help your relationship.

Continue to work toward parity and you will see negative pressures on your marriage diminish. (Remember that equals attract.)

(Turn to Test 2 Scoring Instructions on page 259.)

High Marriage Readiness

The present high divorce rate would drop dramatically if all married persons truly valued marriage as you do. This profile has tremendous strength. When we are ready for marriage, we are much more willing to deal with the problems and pressures of married life. And, most importantly, we will work to improve our relationship because we value it.

A word of caution is in order for those with high marriage readiness. Your readiness makes you more vulnerable to excessive demands and might cause you to compromise more than is good for your relationship. Unless your mate also values your relationship, major problems can develop. And if the relationship doesn't survive, those who value marriage most highly are the ones who suffer most. Because you value marriage so highly, you have every reason to protect it by working to establish and maintain a parity relationship. Be sure that your partner recognizes both your net worth and your intention to add to it.

(Turn to Test 2 Scoring Instructions on page 259.)

Scoring Instructions

VALUES HIERARCHY
Unmated Only

Directions: There are two steps in scoring the Values
Hierarchy Test 2. They must be followed
exactly.

Step 1.   Count the number of A's, B's, C's, etc., in My
Values Column (Column X) on page 235,
and enter the numbers here:
(Example: A __7__ , B __3__ , C __9__)

My Values Scores:

A____   B____   C____   D____   E____
F____   G____   H____   I____   J____

Step 2.   From "My Values Scores" (Step 1 above),
enter the *letter* with the highest number of
responses in space 1 in the "My Values"
column below. Place in space 2 the *letter* with
the next highest number of responses.
Continue until all ten letters are entered in
the "My Values" column.

In case of a tie, when two or more values
receive the same number, place them on the
same line and skip the next number in order.

MY VALUES

1. ____   Example: 1. __C__
2. ____   2. __A__
3. ____   3. __B__
4. ____
5. ____

6. _____   Example for tie:
7. _____   1. <u>A</u>
8. _____   2. <u>I, F</u>
9. _____   3. <u>Leave blank</u>
10. _____   4. <u>B</u>

Step 3. This is your Values Hierarchy. Study the order of Your Marriage Values by comparing your hierarchy with the items which are listed on page 263.

(Unmated Turn to Chapter 12 on page 266.)

Scoring Instructions

VALUES HIERARCHY
Mated Only

Directions: There are eight steps in scoring the Values Hierarchy Test. They must be followed exactly.

Step 1. Count the number of A's, B's, C's, etc., in My Values Column X on page 235, and enter the numbers here: (See example on page 262.)

My Values Score:

A____ B____ C____ D____ E____
F____ G____ H____ I____ J____

Step 2. Count the number of A's, B's, C's, etc. in Mate's Column Y on page 235, and enter the numbers here:

Mate's Values Scores:

A____ B____ C____ D____ E____
F____ G____ H____ I____ J____

Step 3. From "My Values Scores" (Step 1 above),
enter the *letter* with the highest number of
responses in space number 1 in Column X
below. Place in space 2 the *letter* with the next
highest number of responses. Continue until
all ten letters are entered in Column X (See
example below.) Then repeat the same
process for Column Y, transferring the
letters from Step 2 above. In case of a tie,
when two or more values receive the same
number, place them on the same line and
skip the next number in order.

Column X My Values	Column Y Mate's Values	Column Z Discrep- ancy	Example:
			1. __G__
1. ____	1. ____	A. ____	2. __B__
2. ____	2. ____	B. ____	3. __E__
3. ____	3. ____	C. ____	Etc.
4. ____	4. ____	D. ____	
5. ____	5. ____	E. ____	Example
6. ____	6. ____	F. ____	for Tie:
7. ____	7. ____	G. ____	
8. ____	8. ____	H. ____	1. __D__
9. ____	9. ____	I. ____	2. __H,A__
10. ____	10. ____	J. ____	3. (Leave
	TOTAL ____		__blank__)
			4. __I__

Step 4. Now locate and note the letter "A" in Column
X (Step 3) and the letter "A" in Column Y.
Your next job is to subtract the lower number
from the higher number and place the
difference in the Discrepancy Column (Z).

For example, if letter "A" is number 7 in
Column X and number 3 in Column Y,

subtract and place a "4" next to the letter "A" in the Discrepancy Column. If letter "D" were number 2 in Column X and number "9" in Column Y, subtract and place a "7" in the Discrepancy Column. If you have problems with this, look at the sample instructions on page 262.

Step 5. Add the numbers in the Discrepancy Column (Column Z) for your Discrepancy Score and place score here: ________ .

Step 6. Determine the level of your Discrepancy Score by using the following scale and place a (√) by your level.

________	Low Value Agreement	36 - 50
________	Medium Value Agreement	21 - 35
________	High Value Agreement	0 - 20

Step 7. You will find it interesting to examine your values hierarchy as well as your mate's. On page 263 is the list of values included in the test. Match the letters in Column X with the values to determine your values hierarchy. Do the same with the letters in Column Y for your mate's values hierarchy.

Step 8. The final step is to review your profile for your Values Hierarchy. Turn to the appropriate page listed below:

Low Agreement	Page 264
Medium Agreement	Page 264
High Agreement	Page 265

Scoring Instructions

VALUES HIERARCHY
Mated

Step 1. Count the number of A's, B's, C's, etc., in My Values Column (Column X) and enter the numbers here:

My Values:

A __10__ B __5__ C __4__ D __5__ E __3__
F __7__   G __5__ H __1__ I __2__ J __3__

Step 2. Count the number of A's, B's, C's, etc., (as in Item 1) in Mate's Column (Column Y) and enter the numbers here:

A __2__   B __5__ C __3__ D __1__ E __8__
F __10__ G __3__ H __0__ I __4__ J __9__

Step 3. From "My Values Scores" (Step 1 above), enter the *letter* with the highest number of responses in space number 1 in Column X below. Place in space 2 the *letter* with the next highest number of responses. Continue until all ten letters are entered in Column X (See example below.) Then repeat the same process for Column Y, transferring the letters from Step 2 above. In case of a tie, when two or more values receive the same number, place them on the same line and skip the next number in order.

	Column X My Values		Column Y Mate's Values		Column Z Discrep- ancy
1.	A	1.	F	A.	7
2.	F	2.	J	B.	1
3.	B,D,G	3.	E	C.	0
4.		4.	B	D.	6
5.		5.	I	E.	4
6.	C	6.	G,C	F.	1
7.	E,J	7.		G.	3
8.		8.	A	H.	0
9.	I	9.	D	I.	4
10.	H	10.	H	J.	5
			Total		31

ITEMS FROM
VALUES HIERARCHY TEST 2

Directions: Examine your values by matching the letters in your values hierarchy with the statements below.

If mated, do the same with your mate's values hierarchy.

Value

A   Satisfaction in having family, raising children.
B   Reliable companionship within marriage.
C   Feelings of security and safety in marriage.
D   Sexual satisfaction in married life.
E   Sharing of household and daily living tasks.
F   Moral and religious aspects of marriage.
G   Economic advantages of being married.
H   Respect and status from being married.
I   Sharing of affection with one another.
J   Social opportunities of married life.

VALUES AGREEMENT
Mated

Low Values Agreement

You and your mate do not place the same value on the various components of married life. The things which are most important to one of you may have little importance to the other. It is possible for a married couple to have markedly different value priorities and still have a successful marriage, assuming both partners place a high value on marriage itself. But it should be recognized that low values agreement can pose a serious danger to relationships.

Communication is the solution to problems arising from low values agreement. If we respect and trust our partners, we can make allowances for differing values. We need to understand our mates' value hierarchies, however, or friction and confusion are inevitable. Similarly, it is vital that our mates understand our values as well. Effective and ongoing communication in a parity marriage can eliminate much of the danger associated with low values agreement. It not only will help you to understand each other, you are quite likely to move a long way toward each other's values. Keep in mind that values can and should be influenced by rational thought. They rarely are modified by heated arguments.

(Proceed to Chapter 12.)

Medium Values Agreement

A medium level of values agreement can lead to satisfying relationships. You and your mate are often in agreement in terms of the value you attach to the various aspects of married life. The things which are important to you tend to be important to your mate as well.

However, some values have different levels of importance to the two of you. The key to dealing with differing value priorities is ongoing and effective communication. So long as both you and your mate understand and respect each other's values, your relationship should not be damaged by this level of values disagreement. And the more you discuss, the more you are likely to see things in similar ways.

You would probably find it a good use of time to review the marriage values at the front of the chapter and determine which ones are of greater importance to you and which of more importance to your mate.

(Proceed to Chapter 12.)

High Values Agreement

You and your mate are in substantial agreement in terms of your marriage values. This means you tend to attach a similar emphasis to the various aspects of married life. Your relationship should benefit dramatically from this agreement. When couples have similar values they understand each other better and feel supported and affirmed because they agree with each other.

Although values agreement is a powerful plus for your marriage, it does not eliminate the need for parity in net worth. You can have similar values while suffering with an unbalanced relationship. You, like all other couples, must safeguard your marriage from the dangers of non-parity.

(Proceed to Chapter 12.)

12

In The Final Analysis

MATING SHOULD BE, above all, a dynamic process. It involves adaptation to problems and working cooperatively for a higher level of relationship than just being married. Dynamic mating is a lifelong project, worthy of the efforts of two people in love.

The word "dynamic" refers to growth, action, change, and process. In that sense, the scores and profiles in this chapter should be considered as an excellent *starting* point for dynamic mating.

As you add up your total net worth and parity scores and evaluate yourself and your mate or potential mate, keep in mind that the various scores and profiles are changeable. You are not "locked in" to them; they can be improved but they also can deteriorate.

We have provided a wide variety of strategies throughout MATING which can enhance your net worth and your mating success. Additional suggestions are included in this chapter. It is important to review these procedures from time to time as you continue your program of increasing your mating power and improving your chances of mating success.

Important thought and discussion issues are presented with the various charts and profiles. You will find it beneficial to make notes of your important discoveries and decisions. We have provided pages at the back of the book for this purpose. You should not expect to have developed all the answers to all of your own questions, of course. Mark those issues which you'd like to come back to, and give yourself the privilege of a little time to find some excellent solutions. After all, the Dynamic Mating Process truly deserves your care and attention.

Finally, these scores and profiles represent a "base" from which to start. We each have special circumstances and interests which can and should be reflected in our net worth score. You should not hesitate to add in whatever special factors you feel are important to you and your mate.

Because the Dynamic Mating System is designed for both the mated and unmated, separate charts and profiles have been provided.

If you are unmated, continue with Chart 1. If you are mated or have evaluated a potential mate, proceed to page 276.

Your Net Worth Score (Unmated)

To calculate your Net Worth, transfer your Mating Factor Scores from Chapters Six through Ten to Chart 1 below. The page numbers for your scores are provided.

Add your factor scores to find your total Net Worth Score.

Chart 1

MY NET WORTH
Unmated

Factors		*Factor Scores*
Personality	(Page 97, Step 4)	
Appearance	(Page 123, Step 4)	
Economics	(Page 174, Step 4)	
Sex/Affection	(Page 197, Step 4)	
Social Status	(Page 215, Step 4)	

TOTAL NET WORTH SCORE = _______

Your Mating Power

Your net worth represents the mating power that you have at this time. Remember that your factor scores, which add up your net worth, are changeable. They can be improved.

Table 1 below shows you how your total net worth score should be interpreted. Place a (√) by the appropriate level.

Table 1

NET WORTH TO MATING POWER

Total Net Worth Scores		Mating Power Levels	(√)
611 to 750	=	Extremely High	_________
471 to 610	=	High	_________
331 to 470	=	Moderate	_________
191 to 330	=	Limited	_________
50 to 190	=	Low	_________

Your Net Worth Profile

Your Net Worth Profile can help you to visualize your strengths and weaknesses. Use Chart 2 below to plot your five factor scores from Chart 1 into a line graph. A sample chart is provided on the following page.

Chart 2

MY NET WORTH PROFILE
Unmated

Factor Scores	Mating Factors				
	Personality	Appearance	Economics	Sex/Affection	Social
150					
140					
130					
120					
110					
100					
90					
80					
70					
60					
50					
40					
30					
20					
10					

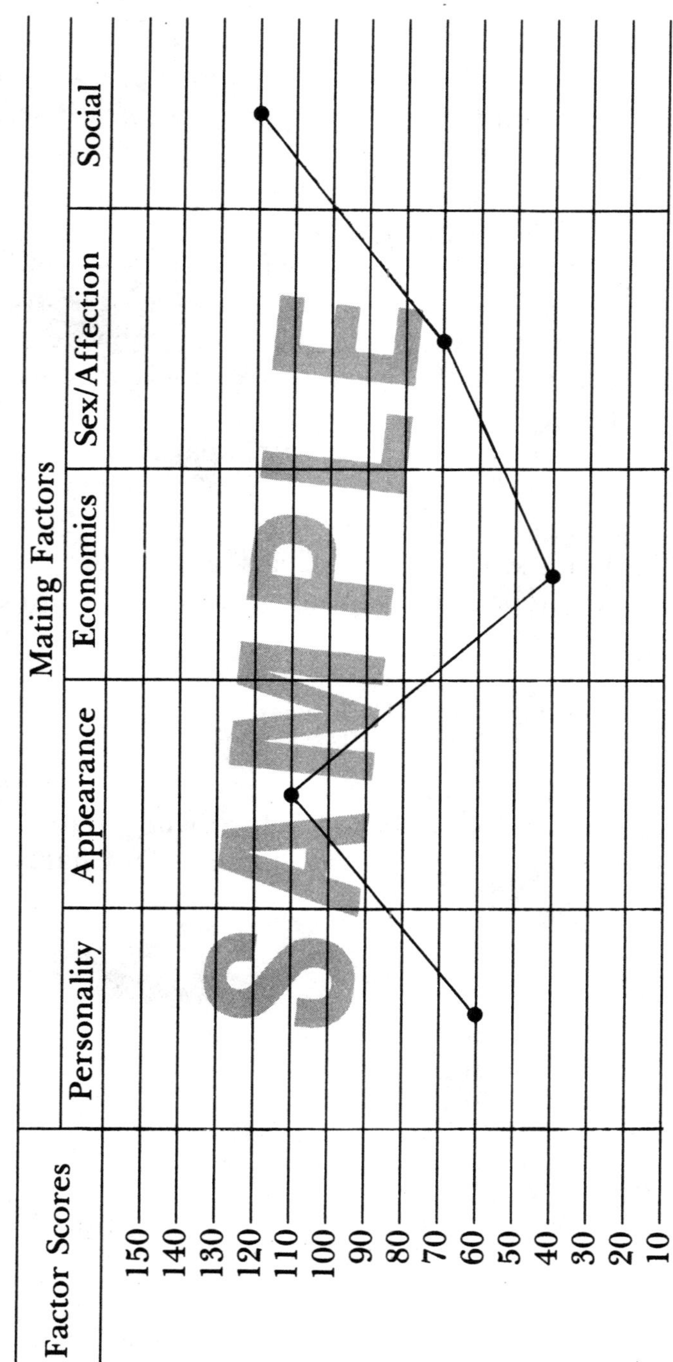

Sample Chart
MY NET WORTH PROFILE
Unmated
Mating Factors
Personality
Appearance
Economics
Sex/Affection
Social
Factor Scores
150
140
130
120
110
100
90
80
70
60
50
40
30
20
10
SAMPLE

Self Analysis Questions

1. Do you have factor scores which are disappointingly low? Which strategies might be most useful to you for improving them? It could be very helpful to you to review the chapters dealing with problem areas.
2. What does your net worth suggest about the desirability of open and closed settings for you?
3. Which of your five mating factor scores suggest strengths which could be further developed for greater mating power? Are there any which might become the kind of speciality discussed in Chapter 3?
4. Would you be happy with a parity mate at your present level? Or would you prefer to increase your net worth to improve your mating power?
5. What kinds of individuals, groups, or organizations would be most helpful in evaluating your net worth?

Special Assets and Liabilities

You may have special assets or liabilities not included in the six mating factors which will influence your net worth significantly. If so, list them in Chart 3 below:

Chart 3

MY ADDITIONAL ASSETS AND LIABILITIES
Unmated Only

Assets

1. ____
2. ____
3. ____
4. ____
5. ____

Liabilities

1. ____
2. ____
3. ____
4. ____
5. ____

It is up to you to decide how much these special factors add to or detract from your mating power. You should use the Net Worth Principle, even informally, when special circumstances influence your assets and liabilities.

Mating Profile

Chart 4 is designed to allow you to compare extremely important test findings in one place. Transfer your factor levels (L, M, or H), your "least acceptable in a mate" factor levels, and "factor importance" levels from the earlier chapters. Page numbers are provided for your convenience.

Then transfer your Marriage Readiness Level and Total Net Worth Level from the pages indicated.

Chart 4

MY MATING PROFILE
Unmated

Mating Factors	My Factor Levels (Step 5)	Least Acceptable In a Mate (Filter)	Importance Level
Personality	(p. 97)	(p. 94)	(p. 95)
Appearance	(p. 123)	(p. 119)	(p. 120)
Economics	(p. 174)	(p. 170)	(p. 171)
Sex/Affection	(p. 197)	(p. 193)	(p. 194)
Social Status	(p. 215)	(p. 211)	(p. 212)

Marriage Readiness Level (p. 249): _______________

Total Net Worth Level Table 1, (p. 269): _______________

Self Analysis Questions

1. Are your own factor levels the same or different from those you want in a mate? Are your expectations, as reflected in your filter levels, appropriate to your strengths and weaknesses?
2. Which factors are most important to you? Are you overlooking any which *should* be important? What kinds of compromises are you willing to make in your selection of a mate?
3. Can you see any significant discrepancies among any of the levels in the three columns?
4. What about your Marriage Readiness score? It should play a determining role in your mating decisions, because if you do not value a marriage, it will not succeed, regardless of your net worth.
5. Given what you want in a mate, where are you most likely to meet and impress candidates for marriage? In open or closed settings?

(Unmated go directly to page 297.)

(Mated Start Here)

Net Worth and Parity

To calculate your Net Worth Score and Parity Score, plus make a few other important comparisons, you need to transfer some scores from Chapters Six through Ten to Chart 5 below. It is provided to make your arithmetic easy and your comparisons clear. Page numbers are provided for your convenience.

After transferring the various scores, simply add your factor scores to find your Total Net Worth Score. Then total your mate's factor scores to determine your mate's Net Worth score.

Chart 5

NET WORTH AND PARITY
Mated

Factors	My Factor Scores	Mate's Factor Scores	Factor Parity
Personality	(p. 99)	(p. 100)	(p. 101)
Appearance	(p. 127)	(p. 127)	(p. 128)
Economics	(p. 177)	(p. 177)	(p. 178)
Sex/Affec-tion	(p. 199)	(p. 200)	(p. 200)
Social Status	(p. 218)	(p. 218)	(p. 219)

Total for Self: _____________ Total for Mate: _________

The Total for Self is *your* Net Worth Score.

The Total for Mate is your *mate's* Net Worth Score

Mating Power

Your Net Worth represents the mating power that you have at this time based on five mating factors. Your mate's mating power is represented by his or her Net Worth Score.

Remember that the factor scores, which make up both Net Worth Scores, are changeable. They can be improved.

Table 2 below shows how your Net Worth Score and your mate's Net Worth Score should be interpreted. Place an "S" by Your Mating Power Level and an "M" by your mate's level.

TABLE 2

NET WORTH TO MATING POWER

Total Net Worth Scores		Mating Power Levels	("S" or "M")
611 to 750	=	Extremely High	________
471 to 610	=	High	________
331 to 470	=	Moderate	________
191 to 330	=	Limited	________
50 to 190	=	Low	________

Net Worth Profiles

Your Net Worth Profile and that of your mate can help you visualize relative strengths and weaknesses. Use Chart 6 below to plot *your* five factor scores from Chart 5 into a line graph. Chart 7 is for your mate's Net Worth Profile. A sample chart is provided between Charts 6 and 7.

Chart 6

MY NET WORTH PROFILE
Mated

Factor Scores	Mating Factors				
	Personality	Appearance	Economics	Sex/Affection	Social
150					
140					
130					
120					
110					
100					
90					
80					
70					
60					
50					
40					
30					
20					
10					

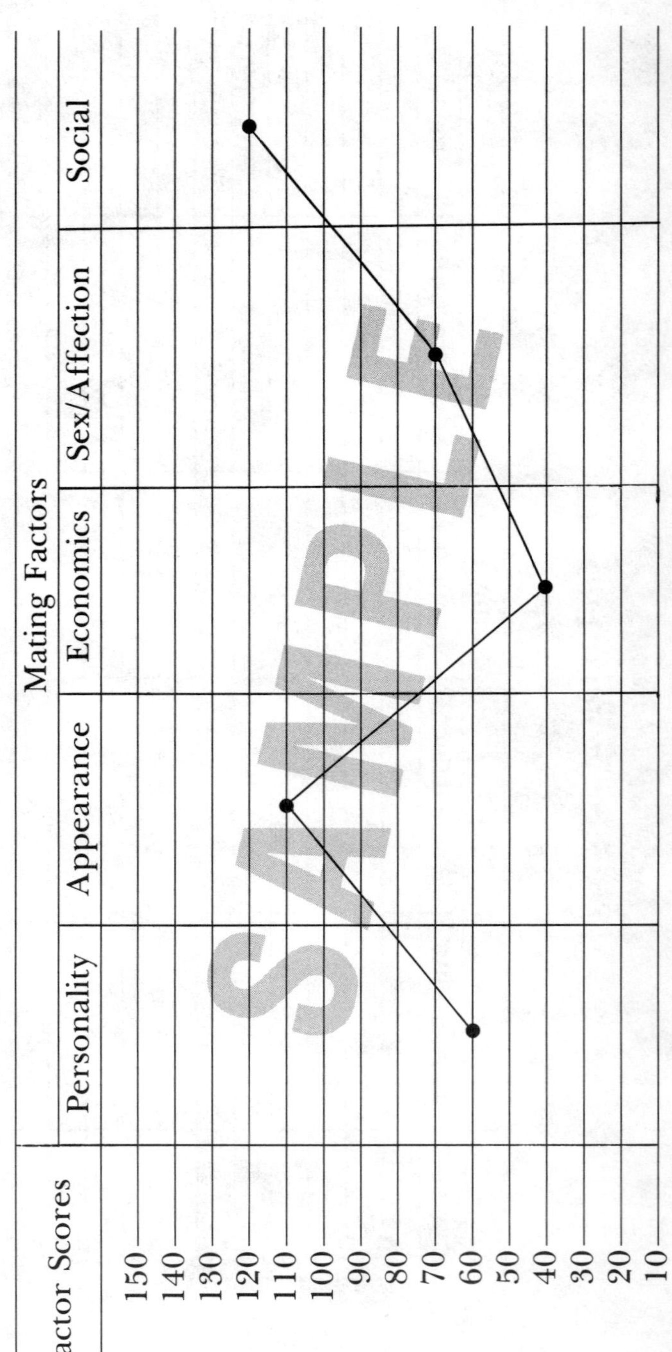
Sample Chart
NET WORTH PROFILE
Mated
Mating Factors
Factor Scores
Personality
Appearance
Economics
Sex/Affection
Social
150
140
130
120
110
100
90
80
70
60
50
40
30
20
10
SAMPLE

Chart 7

MY MATE'S NET WORTH PROFILE
Mated

Factor Scores	Mating Factors				
	Personality	Appearance	Economics	Sex/Affection	Social
150					
140					
130					
120					
110					
100					
90					
80					
70					
60					
50					
40					
30					
20					
10					

Self Analysis Questions

1. Is there a difference between your Net Worth Score and your mate's? We'll discuss parity in the next section.
2. Which factor scores represent your strengths and weaknesses? Your mate's strengths and weaknesses? Are there large discrepancies in some factors?
3. Do you have any factor scores which are disappointingly low? Which strategies might be most useful for improving them? How about your mate's low factor scores?
4. Which of your (or your mate's) five mating factors suggest strengths which could be further developed for greater mating power? Are there any which could become the kind of specialty discussed in Chapter 3?
5. Which kinds of groups or organizations would be helpful in increasing your net worth or your mate's net worth?
6. Are there factors which you and your mate should work on together?

Parity and Net Worth Interaction

Do you and your mate have parity? Subtract the lower number from the higher number to find out. Remember to carry the "S" or "M" (whichever is larger) as you did in earlier chapters. (See the example below.)

Example: If *your* Net Worth Score is 500S and your *mate's* Net Worth Score is 450M, subtract:

$$\begin{array}{r} 500S \\ -450M \\ \hline = 50S \end{array}$$

Before you turn to the parity interpretations we need to add an important note. Both net worth and parity are important to marriage success.

It is not enough to have parity based on low net worth. Even though the partners would have approximate parity, their relative inability to cope with life's problems would put a severe strain on their relationship. It's a major reason that poor people have such high rates of marriage failure. There are exceptions, of course, but we'd like you to have the odds *in your favor* with something as vital as your marriage.

The general rule is this:

> Parity is essential to marriage success. The higher the net worths on which the parity is based, the more likely the marriage is to succeed.

It is useful to think of the following levels of *likelihood for marriage success:*

Parity + High Net Worth = Excellent Prospects
Parity + Medium Net Worth = Moderate Prospects
Parity + Low Net Worth = Limited Prospects
Non-Parity + Any Net Worth Combination = Very
   Poor Prospects

PARITY INTERPRETATIONS FOR "S" SCORES

If your Parity Score ends with "S," your parity interpretation is below.

Excellent Parity

0–170S

Essential parity, but a tendency for you to have a somewhat higher net worth than your mate (unless you have a zero score, in which case you have achieved perfect parity.) If your Marriage Readiness (Chart 7 next) is medium or high and your Values Agreement (Chart 7 next) also is medium or high, you and your mate have achieved a degree of balance in your relationship with excellent prospects for the future. Congratulations! Your challenge will be the joyful one of making a fine relationship even better through mutual growth and support.

Moderate Parity

180S–350S

This parity level represents a fairly well balanced relationship, but one in which your mate is seen as being somewhat deficient in mating power. The relationship should be a healthy one, however, if you have medium or high Marriage Readiness (Chart 7 next) and your Values Agreement (Chart 7 next) also are at those levels. This profile does present a moderate level of risk, however. You have not achieved parity. Your mate is viewed as having lower net worth. It is important to use the strengths· and resources the two of you have to provide for continued growth toward parity. Return to the Net Worth Chart on page 277, and determine in which areas your mate falls below your net worth. Spend additional time with those chapters treating areas in which your mate's net worth is markedly below your own. Determine the specific areas in which there is the greatest discrepancy between you and your mate.

The important point is this: Your relationship is basically sound, but requires attention if you are to achieve parity and avoid the problems which always accompany non-parity marriages. It is important to use the combined resources of your relationship to achieve parity and continue to improve your respective net worths.

Minimal Parity

360S–520S

This relationship is in trouble and you are at risk in your marriage. You view your mate as far below you in net worth. If either your Marriage Readiness Score (Chart 7 next) or Values Agreement Score (Chart 7 next) is at a low level, your relationship is faced with imminent collapse. Steps need to be taken *IMMEDI-ATELY*. First, return to the Net Worth and Parity Chart, page 277. Identify those factors (Personality, Appearance, etc.) which are least similar. Talk to your mate about the disparity. Sessions with a marriage therapist are highly recommended. In conjunction with the therapy and under the supervision of the therapist, encourage your mate to go through the Dynamic Mating System and compare findings. The point is that if you care about your relationship, you will take steps immediately to eliminate the gross disparity which characterizes your marriage. It is important to use the combined resources of your relationship to achieve parity and continue to improve your respective net worths.

Non-Parity
530S–700S

Your marriage is the exception if you are successful. Perhaps you and your mate sustain your relationship for some of the reasons cited in Chapter 2, pages 23–24. At any rate your marriage apparently is seriously stressed by non-parity, but your interest in doing something about it (as reflected in the time you have spent with this book) is a positive sign. If your mate also is interested in improving your relationship, there still may be hope. What to do?

First, your marriage needs help if it is not to fail. Find a competent marriage therapist and begin working on your problems. Second, be very much aware of what research tells us about individuals in your position. When we feel we are grossly superior to our mate, there is a strong tendency to justify lack of commitment to the relationship because we feel that we are being "short-changed." There are forces holding your relationship together at the present time, but the relationship is a deeply troubled one which can only be the source of frustration and discontent until drastic changes are effected.

PARITY INTERPRETATIONS FOR "M" SCORES

If your Parity Score ends with "M," your parity interpretation is below.

Excellent Parity

0–170M

Essential parity has been achieved (unless you have a "zero" score, which is perfect parity). There is a slight tendency, however, for your mate to have a higher net worth than you. If your Marriage Readiness is at a medium or high level (Chart 7 next) and your Marriage Values Score is also at these levels (Chart 7 next), you have a truly outstanding relationship. Congratulations! Your challenge will be the joyful one of making a fine relationship even better through mutual growth and support.

Moderate Parity

180M–350M

This parity level represents a fairly well balanced relationship, but one in which you see your mate as having somewhat higher net worth than you. The relationship should be fairly solid, assuming you are at medium or high levels on both the Marriage Readiness Scale (Chart 7 next) and Values Agreement Test (Chart 7 next). The danger of this score should not be underestimated, however. You have not achieved parity. Your mate is viewed as having higher net worth. We suggest that you think of ways to use the strength and resources of the two of you to provide for continued growth, and for you to make a special effort to catch up.

Return to the Net Worth Chart, page 277, and determine in which areas you fall below your mate in net worth. Then go back to the chapters treating those areas in which you have low net worth. Determine the specific areas in which the discrepancy is greatest between you and your mate. Which assets can you develop for fair trade? Remember, you do not have parity at this time and your relationship is likely being supported by drawing from past savings. While your relationship is basically sound, it does need attention if it is to achieve the balance you both deserve.

Minimal Parity

360M–520M

Your relationship is in trouble and your marriage is at serious risk. You view yourself as markedly below your mate in net worth. If either your Marriage Readiness Score (Chart 7 next) or Values Agreement Score (Chart 7 next) is at a low level, your relationship is faced with imminent collapse. Steps should be taken *IMMEDIATELY*.

First, return to the Net Worth Chart, page 277. Locate those factors (Personality, Appearance, etc.) in which your scores are most dissimilar. What can be done to improve your mating power? Communicate your concern with your mate. Sessions with a marriage therapist are highly recommended. In conjunction with the therapy and under the supervision of the therapist, encourage your mate to go through the Dynamic Mating System and compare findings. The point is this: If you care about saving your relationship, you will take steps immediately to eliminate the serious disparity which characterizes it.

Non-Parity

530M–700M

Your relationship is in desperate condition from a parity standpoint, and if it succeeds, it will be an exception to the rule. Perhaps you and your mate sustain your relationship for some of the reasons cited in Chapter 2, pages 23–24. At any rate, your marriage is seriously stressed by non-parity, but your interest in doing something about it (as reflected in the time you have spent with this book), is a positive sign.

If your mate also is interested in improving the relationship, there still may be hope. What to do?

First, your marriage needs help if it is to survive. Find a competent marriage therapist and begin working on your problems. Second, be very much aware of what research tells us about individuals in your position. People who feel markedly inferior to their mates often are possessive, suspicious and insecure. This, of course, would be neither healthy for you nor your mate. There are current forces holding your relationship together, but the relationship is a deeply troubled one which only can be the source of frustration and discontent until drastic changes are made.

Mating Profile

Chart 7 below is designed to allow you to compare in one place some extremely important Dynamic Mating System findings. Transfer your factor levels (L, M, or H) and your mate's factor levels from the pages indicated. Then transfer your "Importance of Factor" levels from the pages indicated.

Next, *your* Marriage Readiness level, your mate's Marriage Readiness level and the Values Agreement level should be transferred from the pages indicated.

Chart 7

MATING PROFILE
Mated

Factors	My Factor Levels	Impor-tance To Me	Mate's Factor Levels	Impor-tance To Mate
Personality	(p. 100)	(p. 95)	(p. 100)	(p. 95)
Appearance	(p. 127)	(p. 120)	(p. 127)	(p. 120)
Economics	(p. 177)	(p. 171)	(p. 177)	(p. 171)
Sex/ Affection	(p. 199)	(p. 194)	(p. 200)	(p. 194)
Social Status	(p. 218)	(p. 212)	(p. 218)	(p. 212)

My Marriage Readiness Level: (p. 250) ________

My Mate's Marriage Readiness Level: (p. 250) ________

Values Agreement Level: (p. 261) ________

Self Analysis Questions

1. Are there differences between you and your mate in terms of which factors are considered important?
2. What kinds of differences and similarities do you find between your factor levels and your mate's?
3. Do you or does your mate have a low Marriage Readiness level? If so, it can be extremely dangerous for your relationship. If either or both do not value a relationship, it will not succeed. The one with less interest in the marriage has an advantage in debates, because he or she has less to lose if the marriage fails. This too can be extremely disruptive of a relationship. Professional marriage therapy would be appropriate.
4. If your Values Agreement level is low, we recommend that you spend a considerable amount of time with your mate discussing what you value. We will discuss mutual goals and conferences later in this chapter. It is important to agree on values, but as we indicated in Chapter 11, differences can be tolerated and the marriage will not be unduly handicapped.

Special Assets or Liabilities

Your own net worth, or that of your mate, may be influenced by special assets or liabilities. If there are special circumstances in your relationship not included in the six mating factors, list them in Charts 8 and 9 below.

Chart 8

MY ADDITIONAL ASSETS AND LIABILITIES
Mated Only

Assets

1. ____
2. ____
3. ____
4. ____
5. ____

Liabilities

1. ____
2. ____
3. ____
4. ____
5. ____

Chart 8

MY MATE'S ADDITIONAL ASSETS AND LIABILITIES
Mated Only

Assets

1. ____
2. ____
3. ____
4. ____
5. ____

Liabilities

1. ____
2. ____
3. ____
4. ____
5. ____

It is up to you to decide how much these special factors add to or detract from your mating power or your mate's. You should use the Net Worth Principle, even informally, when special circumstances influence assets and liabilities.

MATING STRATEGIES

We've described a great many strategies for building a successful, dynamic relationship. The first and foremost strategy is, as you know, to marry that special someone who has both your love and parity. Your best chance for an excellent marriage, when all is said and done, is to start off with someone you have chosen with your mind as well as your heart.

Dynamic mating, with dynamic love as its centerpiece, requires effort and the trust which comes from mutual experience. It is a relationship which doesn't just happen; it is earned and deserved. Good marriages are caused . . . and so are divorces.

There are four more dynamic mating strategies which deserve our special attention, because they are used in highly successful relationships. For them to be most useful, both spouses need to be aware of them and agree on their importance.

Taking Your Marriage Seriously

Your marriage relationship is the most important social unit you will ever be a part of, and it deserves all the care and attention you can spare. You and your mate should *agree* on this point, however, if you are going to maximize your marital success.

The observations of Marisa and Marvin, an extremely successful young couple, are enlightening:

> "When we started our business together, we poured everything we had into it. Marvin and I were working night and day and it payed off. We had thought it out, and it was a success in no time."
>
> "But our marriage wasn't quite so successful. Marisa was the one who pointed out what we could really do if we took our marriage as seriously as we

took the business. Not that being serious means gloomy. We just wanted more than most others seemed to be having in their relationships. And it's really worked for us."

Mutual Objectives

When two people work toward the same objectives, their mutual purposes and shared efforts have a strengthening effect on their relationship. Bonds of love are increased in a marriage when the same goals are cooperatively sought and the couple shares the challenge and excitement of achievement.

For your relationship to be most successful, you need to define together what success means to you and your spouse.

It helps if you put your goals in writing. Not only will you have a record of what you agreed on, but by writing the goals you are forced to clarify them. Your mate will be helped to understand what you had in mind. For example, "financial success" may be mentioned by both mates as a desirable goal, but it's not clear if it means being retired by the age of 40, becoming a millionaire, enjoying career security, or finally having your credit cards paid off.

Mutual objectives should include goals for yourself, your mate, and the marriage relationship itself.

Mutual goals for yourself?

Yes, because you should be able to expect that your spouse will share those goals for you as an individual and help you achieve them. In the same way, you should plan to help your mate achieve his or her personal objectives.

Objectives for the relationship itself could include owning your own home, enjoying a month of travel each year, producing five children and being able to feed them all, enjoying sex and affection together at least "x" times per week (we suggest revising plans yearly, however), developing a social network of friends, agreeing to tolerate each other's oh-so-wrong political views . . . the list is limited only by your imagination.

Be expansive. Most people's goals are far too modest. And if you define them, they are much more likely to be achieved. They can be personal objectives (becoming more socially effective) or lofty (helping to bring peace to the world), but your marriage success will depend on having them and making them as objective as possible.

One side benefit of having important marriage goals: trivial issues, about which most couple quibble endlessly, are recognized as being relatively unimportant.

Talking and Trading

In order to work together toward mutual goals, you will need to do some talking about obstacles, short cuts, solutions, and successes. And you'll need to do some compromising and trading too, so you will each know who has responsibility to do what.

One of the biggest problems most couples have is insufficient conference time. They usually have the time, they just don't use it for conferences. In fact, most couples do remarkably little talking about anything except the most mundane observations and "Pass the butter, please." According to one source, two-thirds of the couples *avoid* discussions of differences as a way of dealing with problems, and one-tenth *never* discuss problems at all!

How much conference time is necessary? In our experience, one-hour discussions two times per week are *minimal*. The more actively you are pursuing mutual goals, the more excitement and adventure and challenge (and threat of failure) you will have to talk about.

For those couples who are not getting along well, who do not have much parity and whose marriages are suffering, even more time should be spent.

The problem is that it's not fun talking about serious problems, because emotional outbursts and recriminations are frequent. People with serious difficulties should use marriage therapy, and it is quite appropriate (as consultation) for any couple wishing to add more meaning and zest to their relationship.

As couples achieve greater parity, fewer disagreements are going to occur. Happily married people, who have much in common, are going to be able to solve their marital differences through discussion because (a) they are similar, understand each other, and support and trust each other, and (b) their problems usually are not all that serious.

People who rate their marriages as unhappy or average are more likely to report that talking resolves differences only sometimes, seldom, or never. As we indicated, they are the ones who need marriage counseling the most.

It is important to set aside special times, on a regular basis, for your conferences. Select times and places where you will not be interrupted any more than an important corporate business meeting would be. Your marriage is easily as important as any corporation.

Take care not to attempt conferences when you are going to be tired and hungry or distracted too much. One of the authors schedules weekly drives with his family to a variety of places, not so much for the destination as for the family conference which takes place in the van. As his three-year-old daughter Jeanine describes it, "Are we going for another drive-talk, Daddy?"

You might enjoy "drive-talks" too, or prefer the kitchen table. Wherever you do your conferring, it's a time for negotiating and trading and compromising and working to make your marriage successful.

Taking Care of Each Other and Your Team

Finally, your relationship will thrive if you can agree to take care of one another's feelings and emotional needs. This is quite a commitment, so let's take a look at what it involves.

Each of us has the need to be accepted and approved of and loved. We need the security of knowing that we will not be abandoned at the first sign of trouble, that we are appreciated for being valuable and worthwhile. To the extent that we are nurtured and reassured by others, we gain confidence and become more effective in almost every way.

Who is the most important "reassurer" in our lives? Our mates, far and away. They know us best, including our weaknesses and vulnerabilities. If they can approve, and continue to provide love and reassurance, we stand taller and sing louder than ever before.

When we agree, even make a pledge, that we will watch out for one another and provide the ego boosting and reassuring that everyone needs, it gives us a feeling of confidence and security that is truly amazing. Even when you know that the statement of love is a prearranged gesture, it still works. It works because it is a gift.

Try this experiment. Ask your mate or potential mate to say to you, "I love you darling, and I always will."

It likely will make you feel wonderful, even though you asked for it!

Incidentally, if your mate refuses to make the statement, start again on page one of this book.

The point to remember is that the agreement to take care of each other is a conscious agreement, not something you can take for granted. We need a lot more "I love you's" in our lives than most people would imagine.

The two of you form a team, and as with any team, it is more than the sum of its parts. That is, the "marriage team" is an entity, in and of itself, like a baseball team, or a partnership or a company. We won't go so far as to say your team deserves a special uniform or logo or mascot, but it certainly warrants special attention.

By thinking of your marriage as a team, with shared experiences and goals and insights, it becomes, in a very real sense, a living entity, like the family you were part of as a child. And that living unit deserves your care and attention, so you will be able to take pride not only in its accomplishments, but in its nature and style and the way it functions. It is, after all, a part of you.

Footnotes

Chapter One

1. Cherlin, A. J., *Marriage, Divorce, Remarriage.* Cambridge, Massachusetts: Harvard University Press, 1981, p. 69.
2. *U. S. News and World Report*, July 5, 1982, p. 10.
3. Ibid.
4. Cherlin, p. 22.
5. Wallerstein, J. S. and J. B. Kelly, The effects of parental divorce: The adolescent experience. In E. J. Anthony and C. Koupernik (Eds.), *The Child in His Family* (Vol. 3), New York: Wiley, 1974.
6. Wallerstein, J. S. and J. B. Kelly, The effects of parental divorce: Experiences of the pre-school child. *Journal of the American Academy of Child Psychiatry*, 1975, *14*, 600–616.
7. Kulka, R. A. and H. Weingarten, The long term effects of parental divorce in childhood on adult adjustment. *Journal of Social Issues, 35*(4), 1979, p. 51.
8. Ibid., p. 68.
9. Cherlin, p. 90.
10. Duncan, D. and T. R. Duncan, Your guide to splitting. *Harper's Bazaar*, Jan. 1980, p. 97.
11. Espenshade, T. J., The economic consequences of divorce. *Journal of Marriage and the Family*, August 1979, p. 617.
12. Kinsey, A. C., W. Pomeroy, and C. E. Martin. *Sexual Behavior in the Human Male*. Philadelphia: Saunders, 1948.

13. Kinsey, A. C., W. Pomeroy, C. E. Martin, and P. H. Gebhard. *Sexual Behavior in the Human Female*. Philadelphia: Saunders, 1953.
14. Johnson, R. E., Some correlates of extramarital coitus. *Journal of Marriage and the Family*, August 1970, pp. 449 and 454.
15. Johnson, R. E., Attitudes toward extramarital relationships, *Medical Aspects of Human Sexuality*, 1972, 6(4), 168–191.
16. Levinger, G., A social psychological perspective on marital dissolution, *The Journal of Social Issues*, 1976, 21–48.

Chapter Two

1. Proxmire, W. Quoted in *Time Magazine*, March 24, 1975.
2. The New Scarlet Letter. *Time Magazine*, August 2, 1982, p. 62.
3. *New York Magazine*, August 16, 1982, p. 9.
4. *Statistical Abstracts of the U. S., 1981. National Data Book and Guide to Sources*, Washington, D. C.: U. S. Bureau of the Census.
5. Cherlin, p. 5.
6. Kurian, G. T. *The Book of World Rankings*, New York: Facts on File, 1979, p. 32.
7. Albert, G., *Choosing and Keeping a Marriage Partner*. Berea, Ohio: Personal Growth Press, 1971.

Chapter Three

1. Walster, E. and G. W. Walster, *A New Look at Love*. Reading, Massachusetts: Addison Wesley Publishing Co., 1978, p. 134.
2. Huston, T. L., A decision-making model of the initiation of heterosexual pair relationships. Paper Presented at the National Council in Family Relations, Toronto, October 1973.
3. Ibid.
4. Walster, E., G. W. Walster, J. Piliavin, and L. Schmidt, Playing hard-to-get: understanding an elusive phenom-

enon. *Journal of Personality and Social Psychology*, 1973, *26*, p. 120.

5. Walster, E., V. Aronson, D. Abrahams and L. Rottman, Importance of physical attractiveness in dating behavior, *Journal of Personality and Social Psychology*, 1966, *5*, p. 508.

6. Walster and Walster, p. 135.

7. Walster, E., J. Traupman and G. W. Walster. Equity and extramarital sexuality. *Archives of Sexual Behavior,* 1978, *7*, 127–141.

8. *Teenage pregnancy*. Washington, D. C.: U. S. Government Printing Office, 1977.

9. Pharis, M. E. and Martin Manosevitz, Parental models of infancy: A note on gender preferences for firstborns. *Psychological Reports*, 1980, *47*, p. 763.

10. Murstein, B. I. *Who Will Marry Whom?* New York: Springer Publishing Company, 1976, p. 168.

11. Etzioni, A. Sex control, science and society. *Science*, 161, p. 1109.

12. Pearce, D., How long till equality? *Time Magazine*, July 13, 1982, p. 24.

13. Ibid., p. 23.

14. Murstein, pp. 168 and 177.

15. Ibid.

16. Ibid.

17. *New York Magazine*, August 16, 1982, p. 93.

18. *Statistical Abstracts*, 1981.

19. Ibid.

20. Gagnon, J. H., *Human Sexualities.* Glenview, Illinois: Scott Foresman and Company, 1977, pp. 253–4.

21. Kipnis, D., Changes in self-concepts in relation to perceptions of others. *Journal of Personality*, 1961, *29*, 449–465.

22. Murstein, p. 202.

Chapter Four

1. Walster and Walster, p. 45.

2. Tennov, D., *Love and Limerance*. New York: Stein and Day, 1979, p. 142.

3. Reik, T., *A Psychologist Looks at Love*. New York: Holt,

Rinehart and Winston, 1944, p. 150.
4. Rougemont, D., *Love in the Western World*. New York: Pantheon, 1956.
5. Driscoll, R., K. Davis and M. E. Lipetz, Parental interference and romantic love: The Romeo and Juliet effect. *Journal of Personality and Social Psychology*, 1972, *24*, p. 8.
6. Walster and Walster, p. 41.
7. Lindsey, B. and W. Evans., *The Companionate Marriage*. New York: Boni and Liveright, 1927.

Chapter Six

1. Burgess, E. and P. Wallin, *Engagement and Marriage*. Philadelphia: Lippincott, 1953, p. 202.
2. Strauss, A., The influence of parent images upon marital choice. *American Sociological Review, 11*(5), 554–559.
3. Winch, R. F., *Mate Selection: A Study of Complementary Needs*. New York: Harper and Row, 1958.
4. Murstein, B. I., Mate selection in the 1970's. *Journal of Marriage and the Family*, November 1980, 777–792.
5. Murstein, B. I., The relationship of mental health to marital choice and courtship progress. *Journal of Marriage and the Family*, 1967, *26*, 689–696.
6. Lott, B., A possible role for generally adaptive features in mate selection and sexual stimulation. *Psychological Reports*, 1979, *45*.
7. Murstein, The relationship of mental health to marital choice and courtship progress, 689–696.

Chapter Seven

1. Murstein, *Who Will Marry Whom?* p. 168.
2. Elder, G. H. Jr., Appearance and education in marriage mobility. *American Sociological Review*, 1969, *34*, 519–533.
3. Walster and Walster, p. 139.
4. Dion, K. and E. Berscheid, Physical attractiveness and sociometric choice in nursery school children, 1971.

5. Berscheid, E., E. Walster and G. Bohrnstedt, Body image. *Psychology Today*, July 1972, p. 122.
6. Berscheid, E., K. Dion, and E. Walster, What is beautiful is good. *Journal of Personality and Social Psychology*, *24* (3), 285–290.
7. Walster, E., V. Aronson and L. Rottman, Importance of physical attractiveness in dating behavior. *Journal of Personality and Social Psychology*, 1966, *4*, 506–516.
8. Ibid.
9. Bosard, J. H. S., Residential propinquity as a factor in mate selection. *American Journal of Sociology*, *38*, 219–224.
10. Berscheid, E., and E. Walster, Physical attractiveness, in L. Berkowitz (Ed.) *Advances in Experimental Social Psychology*, New York: Academic Press, 1974, 157–215.
11. Berscheid, Walster and Bohrnstedt, p. 210.
12. Huston, T. L., A decision making model of the initiation of heterosexual pair relationships. Paper Presented at the National Council in Family Relations, Toronto, October 1973.

Chapter Eight

1. Espenshade, T. J., The economic consequences of divorce. *Journal of Marriage and the Family*, August 1979, p. 616.
2. Becker, G. S., E. M. Landis and R. T. Michael, An economic analysis of marital instability. *Journal of Political Economy*, 1977, *85*, p. 1161.
3. Landis, J. T. and M. G. Landis. *Building a Successful Marriage*. Englewood Cliffs, New Jersey: Prentice Hall Inc., 1977, p. 317.
4. Blood, R. O. Jr., *Anticipating Your Marriage*. New York: The Free Press of Glencoe, 1962, p. 292.
5. Wilson, P. P., College women who express futility. *Contributions to Education*, New York: Columbia University, 1950, p. 54.
6. Pace, R., *They Went to College*. Minneapolis: University of Minnesota Press, 1941, p. 82.

7. Population Reference Bureau, U. S. women at work. *Population Bulletin 36*(2), p. 29.
8. Murstein, B. I., *Love, Sex and Marriage Through the Ages.* New York: Springer Publishing Co., 1974, p. 48.
9. Ibid., p. 64.
10. Ibid., p. 126.
11. Ibid., p. 259.
12. Kerckhoff, A. C., Status-related value patterns among married couples. *Journal of Marriage and the Family*, February, 1972, p. 105.
13. Murstein, *Love, Sex and Marriage Through the Ages*, p. 406.
14. Terman, L. M., *Psychological Factors in Marital Happiness.* New York: McGraw-Hill Book Company, Inc., 1938, 167–171.

Chapter Nine

1. Murstein, *Who Will Marry Whom*, p. 226.
2. Murstein, *Love, Sex and Marriage Through the Ages*, p. 406.
3. Wolfe, L., *The Cosmo Report*. New York: Arbor House, 1981, p. 84.

Chapter Ten

1. Murstein, *Love, Sex and Marriage Through the Ages*, p. 406.
2. Cavan, R. S., *The American Family* (2nd Ed.). New York: Thomas Y. Crowell Company, 1953, p. 232.
3. LeMasters, E. E., *Modern Courtship and Marriage*. New York: The Macmillan Company, 1957, p. 285.

Chapter Eleven

1. Newcomb, T. M., *The Acquaintance Process*. New York: Holt, Rinehart and Winston, 1961.
2. Walster, E. and G. W. Walster, Effect of expecting to be liked on choice of associates. *Journal of Abnormal and Social Psychology*, 1963, *67*, 402–404.

3. Murstein, *Love, Sex and Marriage Through the Ages*, p. 406.
4. Byrne, D. and G. L. Clore, Jr., Predicting interpersonal attraction toward strangers in three different stimulus modes. *Psychological Science*, 1966, *4*, 239–240.
5. Coombs, R. H., Value consensus and partner satisfaction among dating couples. *Journal of Marriage and the Family*, May 1966, p. 165.
6. Byrne, D., *The Attraction Paradigm*. New York: Academic Press, 1971.
7. Clore, G. and D. Byrne, A reinforcement-affect model of attraction, in T. L. Huston (Ed.) *Foundations of Interpersonal Attractions*, New York: Academic Press, 1974, pp. 143–170.
8. Medling, J. A. and Michael McCarrey, Marital adjustment over segments of the family life: the issue of spouses value similarity. *Journal of Marriage and the Family*, February 1981, p. 195.

SIDEBAR FOOTNOTES

1a. Cherlin, A. J., *Marriage, Divorce, Remarriage*. Cambridge: Harvard University Press, 1981, pp. 23 and 69.
2a. Adapted from U.N. Demographic Yearbook, in G. T. Kurian, *The Book of World Rankings*, New York: Facts on File, 1979, p. 32.
3a. Gettings, F., *The Book of the Hand: An Illustrated History of Palmistry*. London: The Hamlyn Publishing Group Ltd., 1965, pp. 140–2 and 107–8.
4a. Morris, S., *The Book of Strange Facts and Useless Information*. Garden City: Doubleday and Company, Inc. (Dolphin Books), 1979, p. 82.
5a. McWhirter, N., *Guiness Book of World Records, 1980 Edition*. New York: Sterling Publishing Company, Inc., 1979, p. 449.
6a. Wallace, I., D. Wallachinsky, A. Wallace and S. Wallace, *The Book of Lists #2*. New York: William Morrow and Company, Inc., 1980, p. 362.
7a. Weiser, M. P. K. and J. S. Arbeiter, *Womanlist*. New York: Atheneum, 1981, p. 53.

8a. Linton, R. in V. Packard, *The Sexual Wilderness*. New York: David McKay Company, Inc., 1968, p. 226.

9a. Flexner, S. B., *I Hear America Talking*. New York: Van Nostrand Reinhold Company, 1976, 101–2.

10a. Folson, J. K., The romantic complex and cardiac respiratory love, *The Family*, New York: Wiley, 1934, p. 356.

11a. Mudd, E. in V. Packard, *The Sexual Wilderness*, p. 484.

12a. *Standard and Poor's Industry Surveys*, May 20, 1982. Section 2, p. H27.

13a. Weiser and Arbeiter, p. 269.

Bibliography

Adams, B. N. *The American Family: A Sociological Interpretation*. Chicago: Martham, 1972.

Albert, G. *Choosing and Keeping a Marriage Partner*. Berea, Ohio: Personal Growth Press, 1971.

Altman, I. Reciprocity of interpersonal exchange. *Journal for the Theory of Social Behavior*, October 1973, 249–261.

Altman, I., and Taylor, D. A. *Social Penetration: The Development of Interpersonal Relationships*. New York: Holt, Rinehart and Winston, 1973.

Andelin, H. B. *Fascinating Womanhood*. Santa Barbara, Ca.: Pacific Press, 1963.

Antonucci, T. Attachment: a life-span concept. *Hum. Dev.* 1976, *19*, 135–142.

Aronson, E. *The Social Animal*. San Francisco, California: W. H. Freeman and Co., 1972.

Aronson, E. Some antecedents of interpersonal attraction. In W. J. Arnold and D. Levine (Eds.), *Nebraska Symposium on Motivation* 1969. Nebraska: University of Nebraska Press, 1970.

Azrin, N. H., Naster, B. J., and Jones, R. Reciprocity counseling: a rapid learning-based procedure for marital counseling. *Behavior Research and Therapy*, 1973, *11*, 365–382.

Bachrach, L. L. Marital status and mental disorder: an analytical review. DHEW Publication, Washington, D.C.: U.S. Government Printing Office, 1975, 75–217.

Bailyn, L. Career and family orientations of husbands and wives in relation to marital happiness. *Human Relations*, April 1970, *23*, 97–113.

Bane, M. J. Marital disruption and the lives of children. *Journal of Social Issues*, 1976, *32*, 103–117.

Banta, T. J., and M. Hetherington, Relations between needs of friends and fiances. *Journal of Abnormal and Social Psychology*, *66*, 401–404.

Barbach, L. *For Yourself — The Fulfillment of Female Sexuality.* New York: Doubleday, 1975.

Bardwick, J. *The Psychology of Women: a Study of Biocultural Conflicts.* New York: Harper and Row, 1971.

Barry, W. Marriage research and conflict: An integrative review. *Psychological Bulletin*, 1970, *73*, 41–54.

Barton, K. and Cattell, R. B. Real and perceived similarities in personality between spouses: test of "likeness" versus "completeness" theories. *Psychological Reports*, 1972, *31*, 15–18.

Bayer, A. E. Sexist students in American colleges: a descriptive note. *Journal of Marriage and the Family*, 1975, *37*, 391–397.

Bebbington, A. C. The function of stress in the establishment of the dual career family. *Journal of Marriage and the Family*, August 1973, *35*, 530–537.

Beck, G. D. Person perception and marital adjustment. Master's Thesis. Connecticut College, 1970.

Becker, G. A theory of marriage: Part I. *Journal of Political Economy*, July, August 1973, 813–846.

Becker, G. A theory of marriage: Part II. *Journal of Political Economy*. March, April 1974, *82*, 511–526.

Becker, G. S., Landes, E. M., and Michael, R. T. An economic analysis of marital instability. *Journal of Political Economy 85*, 1977, 1141–87.

Beckman, L. Assortive mating in man. *The Eugenics Review*, 1962, *54*, 63–67.

Beier, E. G., Rossi, A. M., Garfield, R. L. Similarity plus dissimilarity of personality. Basis for friendship? *Psychological Reports*, 1961, *8*, 3–8.

Benson, P. The common interests myth in marriage. *Social Problems*, 1955, *3*, 27–34.

Bentler, P. M. and Newcomb, Michael, D. Longitudinal study of marital success and failure. *Journal of Consulting and Clinical Psychology*, 1978, *46*(5) 1053–1070.

Bernard, J. *The Future of Marriage*. New York: Bantam Books, 1972.

Bernard, J. The adjustment of married mates. In H. T. Christensen, (Ed.), *Handbook of Marriage and the Family*. Chicago: Rand McNally, 1964, 675–739.

Bersheid, E., Dion, K., Walster, C., and Walster, G. W. Physical attractiveness and dating choice: A test of the matching hypothesis. *Journal of Experimental Social Psychology*, 1971, *7*, 173–189.

Bersheid, E., and Fei, J. Romantic love and sexual jealousy. In G. Clanton and Lynn G. Smith, (Eds.), *Jealousy*. Englewood Cliffs, N. J.: Prentice Hall, 1977, 101–109.

Bersheid, E. K., Dion, K., and Walster, E. What is beautiful is good. *Journal of Personality and Social Psychology*, 1972, Vol. 24, No. 3, 285–290.

Bersheid, E., and Walster, E. A little bit about love. In T. L. Huston (Ed.), *Foundations of Interpersonal Attraction*. New York: Academic Press, 1974, 355–381.

Bersheid, E., and Walster, E. Beauty and the best. *Psychology Today*, March 1972, *5*, 42.

Bersheid, E., and Walster, E. Physical attractiveness. In Leonard Berkowitz (Ed.), *Advances in Experimental Social Psychology*. (Vol. 7). New York: Academic Press, 1974, 158–216.

Bersheid, E., and Walster, E. *Interpersonal Attraction* (2nd ed.). Reading, Massachusetts: Addison-Wesley, 1978.

Bersheid, E., Walster, E., and Bohrnstedt, G. Body image. *Psychology Today*, November 1973.

Bersheid, E., Stephen, N., and Walster, E. Sexual arousal and heterosexual perception. *Journal of Personality and Social Psychology*, 1971, *20*, 93–101.

Birtchnell, J. and Mayhew, J. Thomas Theory: Tested for mate selection and friendship formation. *Journal of Individual Psychology*. 1977, *33*, 18–36.

Black, H., and Angelis, V. B. Interpersonal attraction: an emperical investigation of platonic and romantic love. *Psychological Reports*, 1974, *34*, 1243–1246.

Blau, P. M. *Exchange and Power in Social Life*. New York: John Wiley, 1964.

Blazer, J. A. Complementary needs and marital happiness.

Marriage and Family Living, 1963, *25*, 89–95.

Blood, R. O., Jr. *Anticipating Your Marriage*. The Free Press of Glenco, New York: The Macmillan Co., 1962, p. 299.

Blood, R. O., Jr. The husband-wife relationship. In F. I. Nye and I. Hoffman (Eds.) *The Employed Mother in America*. Westport, Connecticut: Greenwood Press, 1976, 282–305.

Blood, R. O., Jr., and Wolfe, D. M. *Husbands and Wives: The Dynamics of Married Living*. Glencoe, Ill.: The Free Press, 1960.

Bloom, B. L., Asher, S. J., and White, S. W. Marital disruption as a stressor: A review and analysis. *Psychological Bulletin*, 1978, *85*, 867–894.

Boalt, Gumnar. *Family and Marriage*. David McKay Co., New York, 1965.

Boeverman, C. E. and Day, B. R. A test of the theory of complementary needs as applied to couples during courtship. *American Sociological Review*, October 1956, *21*, 602–605.

Bolton, C. D. Mate selection as the development of a relationship. *Marriage and Family Living*, 1961, *23* 234–240.

Booth, A. Wife's employment and husband's stress: A replication and refutation. *Journal of Marriage and the Family*, November 1977, *39*, 645–650.

Booth, A. and White, L. Thinking about divorce. *Journal of Marriage and the Family*, August 1980.

Bosard, J. H. S. Residential propinquity as a factor in mate selection. *American Journal of Sociology*, 1932, *38*, 219–224.

Boukydis, C. F. Z. Adult perception of infant appearance: A review. *Child Psychiatry and Human Development*, Summer 1981, Vol. II, (4).

Bowlby, J. Affectional bonds: their nature and origin. In R. W. Weiss, *Loneliness: The Experience of Emotional and Social Isolation*. Cambridge, Mass.: MIT Press, 1973.

Bowman, C. C. Uncomplimentary remarks on complementary needs. *American Sociological Review*, 1955, *20*, 466.

Bowman, T., Giuliani, G., Mingé, M. R. *Finding Your Best Place to Live in America*. West Babylon, N.Y.: Red Lion Books, 1981.

Boxer, L. Mate selection and emotional disorder. *Family Coordinator*, 1970, *19*, 173–179.

Bradburn, N., and Copolvitz, D. *Reports on Happiness*. Chicago: Aldine Publishing Company, 1965.

Braiker, H. B. and Kelley, H. H. Conflict in the development of close relationship. In R. L. Burgess and T. L. Huston (Eds.), *Social Exchange in Developing Relationships*. New York: Academic Press, 1979, 135–168.

Brandwein, R. A., Brown, C. A. and Fox, E. M. Women and children last: The social situation of divorced mothers and their families. *Journal of Marriage and the Family*, 1974, *36*, 498–514.

Brinkerhoff, D. B. and White, L. K. Marital satisfaction in economically marginal population. *Journal of Marriage and the Family*, May 1978, 259–267.

Briscoe, G., Smith, J., Robbins, E., Martin, S., and Gaskin, F. Divorce and psychiatric disease. *Archives of General Psychiatry*. July 1973, *29*, 119–125.

Brislin, R. W., and Lewis, S. A. Dating and physical attractiveness: a replication. *Psychological Reports*, 1968, *22*, 976.

Burchinal, L. B. and Chancellor, L. E. Survival rates among religiously homogamous and interreligious marriages. *Social Forces*, May 1963, *41*, 353–362.

Burgess, E. W. and Cottell, L. S. *Predicting Success or Failure in Marriage*. New York: Prentice Hall, 1939.

Burgess, E. W., and Wallin, P. Homogamy in personality characteristics. *Journal of Abnormal and Social Psychology*, 1944, *29*, 475–481.

Burgess, E. W., and Wallin, P. *Engagement and Marriage*. Philadelphia: Lippincott, 1953.

Burke, R. J. and Tamara, W. Relationship of wives' employment status to husband, wife and pair satisfaction and performance. *Journal of Marriage and the Family*, May 1976, *38*, 279–287.

Butler, R. T. *Beliefs in the Causes of Happiness*. Unpublished doctoral dissertation, Memphis State University, 1974.

Butler, R., and Lewis, M. *Aging and Mental Health: Positive Psychological Approaches*. St. Louis, Missouri: Mosby Co., 1973.

Byrne, D. *The Attraction Paradigm*. New York: Academic Press, 1971.

Byrne, D. Interpersonal attraction and attitude similarity. *Journal of Abnormal Social Psychology*, 1961, *62*, 713–715.

Byrne, D. Cherry, F., Lambert, J., and Mitchell, H. Husband-wife similarity in response to erotic stimuli. *Journal of Personality*, July 1973, *41*, 385–394.

Byrne, D. and Clore, G. L. Jr. Predicting interpersonal attraction toward strangers in three different stimulus modes. *Psychological Science*, 1966, *4*, 239–240.

Campbell, A. The American way of mating. *Psychology Today*, 1975, *8*(1), 37–43.

Campbell, A. Women at home and at work. In D. McGuigan (Ed.), *New Research on Women and Sex Roles*. Ann Arbor, Michigan: Center for Continuing Education of Women, 1976.

Campbell, A., Converse, P. E., and Rodgers, W. L. *The Quality of American Life*. New York: Russell Sage Foundation, 1976.

Carns, D. E. Talking about sex: Notes on first coitus and the double sexual standard. *Journal of Marriage and the Family*, November 1973, *35*, 677–688.

Carter, H., and Glick, P. C. *Marriage and Divorce: A Social and Economic Study* (Rev. ed.) Cambridge: Harvard University Press, 1976.

Cattell, R. B. and Nesselroade, J. B. Likeness and completeness theories examined by sixteen personality factor measures on stably and unstably married couples. *Journal of Personality and Social Psychology*, 1967, *7*, 351–361.

Cavan, R. S. *The American Family*, 2nd ed. New York: Thomas Y. Crowell, 1953.

Center, R. *Sexual Attraction and Love: An Instrumental Theory*. Springfield, Illinois: C. C. Thomas, Publisher, 1975.

Centers, R. and Granville, A. C. Reciprocal need gratification in intersexual attraction: A test of the hypotheses of Shutz and Winch. *Journal of Personality*, 1971, *39*, 26–43.

Cerf, B. *Bennett Cerf's Bumper Crop of Anecdotes and Stories Mostly Humorous About the Famous and Near Famous*. (2 vols.) Garden City, New York: Garden City Books, Inc.

Cherlin, A. *Marriage, Divorce, Remarriage*, Cambridge: Harvard University Press, 1981.

Cherlin, A. The effects of children on marriage dissolution. *Demography*, 1977, *14*, 265–272.

Cherlin, A. and Reeder, L. G. The dimensions of psychological well-being. *Sociological Methods and Research*, November 1975, *4*, 189–214.

Chester, R. Is there a relationship between childlessness and marriage breakdown? *Journal of Biological Science*, 1972, *4*, 443–454.

Christensen, H. T. Children in the family: Relationship of number and spacing to marital success. *Journal of Marriage and the Family*, May 1968, *30*, 283–289.

Cimbalo, R. S., Faling, V., and Mousaw, P. The course of love: a cross sectional design. *Psychological Reports*, 1976, *38*, 1292–1294.

Clark, M. S. and Mills, J. Interpersonal attraction in exchange and communal relationships. *Journal of Personality and Social Psychology*, 1979, *37*, 12–24.

Clore, G. L. and Byrne, D. A reinforcement-affect model of attraction. In T. Houston (Ed.), *Foundations of Interpersonal Attraction*. New York: Academic Press, 1974, 143–170.

Condie, Spencer, and Doan, H. T. Do husbands and wives belong to the same marriage? A multiple role analysis of marital satisfaction throughout the family life cycles. Paper Presented at the Annual Meeting of the Pacific Sociological Association, San Diego, California, March 1976.

Coomb, L. C., Freedman, R., Freedman, J., and Pratt, W. F. Premarital pregnancy and status before and after marriage. *American Journal of Sociology*, 1970, *75*, 800–820.

Coombs, R. H. A value theory of mate selection. *Family Life Coordinator*, 1961, *10*, 51–54.

Coombs, R. H. Reinforcement values in the parental home as a factor in mate selection. *Journal of Marriage and Family Living*, 1962, *24*, 155–157.

Coombs, R. H. Value consensus and partner satisfaction among dating couples. *Journal of Marriage and the Family*, May 1966, 166–173.

Corsini, R. J. Understanding and similarity in marriage. *Journal of Abnormal and Social Psychology*. 1956, *52*, 327–332.

Corson, R. *Fashions in Makeup — from Ancient to Modern Times*. New York: Universe Books, 1972.

Croake, J. W., and James, B. A four year comparison of premarital sexual attitudes. *Journal of Sex Research*, May 1973, *9*, 91–96.

Cromwell, R. E., Olson, D. H., and Fournie, D. G. Diagnosis evaluation in marital and family counseling. In D. H. Olson (Ed.), *Treating Relationships*, Lake Mills, Iowa: Graphic Publishing, 1976, 517–562.

Davis, K. B. *Factors in the Sex Life of Twenty-two Hundred Women*. New York: Harpen, 1929.

Davis, M. S. *Intimate Relations*. New York: The Free Press, 1973.

DeLora, J. R. and DeLora, J. S. *Intimate Lifestyles: Marriage and its Alternatives (2nd ed.)*. Pacific Palisades, California: Goodyear, 1975.

Derlega, W. J. and Chaikin, A. L. *Sharing Intimacy: What We Reveal to Others and Why*. Englewood Cliffs, New Jersey: Prentice Hall, 1975.

Derlega, W. J., and Chaikin, A. L. Norms affecting self-disclosure in men and women. *Journal of Consulting and Clinical Psychology*, June 1976, *44*, 376–380.

DeYoung, G. E., and Fleischer, B. Motivational and personality trait relationships in mate selection. *Behavior Genetics*, 1976, *6*, 1–6.

Dion, K., and Bersheid, E. Physical attractiveness and sociometric choice in nursery school children. Mimeographed Research Report, 1971.

Dion, K. K., and Dion, K. L. Correlates of romantic love, *Journal of Consulting and Clinical Psychology*, 1973, *4*, 41–56.

Dion, K. K., and Dion, K. L. Self-esteem and romantic love. *Journal of Personality*, March 1975, *43*, 39–57.

Dion, K. K., and Dion, K. L. Love, liking and trust in heterosexual relationships. *Personality and Social Psychology Bulletin*, Spring 1976, *2*, 187–190.

Driscoll, R. K., and Davis, K. E. and Lipetz, M. E. Parental interference and romantic love. *Journal of Personality and Social Psychology*, 1972, *24*, 1–10.

Duncan, D., and Duncan, J. R. Your guide to splitting. *Harper's Bazaar*, January 1980.

Dunner, D. L. Assortive mating in primary affective disorders. *Biological Psychiatry*, 1976, *11*, 43–51.

Dutton, D. G. and Aron, A. P. Some evidence for heightened sexual attraction under conditions of high anxiety. *Journal of Personality and Social Psychology*, 1974, *30*, 510–517.

Eckland, B. K. Theories of mate selection. *Eugenics Quarterly*, 1968, *15*, 71–84.

Edwards, J., and Booth, A. Sexual behavior in and out of marriages: An assessment of correlation. *Journal of Marriage and the Family*, February 1976, *38*, 73–81.

Ehrmann, W. *Premarital Dating Behavior*. New York: Bantam, 1960.

Eiderberg, L. Neurotic choice of mate. In V. W. Eisenstein (Ed.) *Neurotic Interaction in Marriage*. New York: Basic Books, 1956, 57–64.

Elder, G. H., Jr. Appearance and education in marriage mobility. *American Sociological Review*, 1969, *34*, 519–533.

Espenshade, T. J. The economic consequences of divorce. *Journal of Marriage and the Family*, August 1979, 615–625.

Etzioni, A. Sex control, science and society. *Science*, 1968, 161.

Eysenck, H. J. Personality, premarital sexual permissiveness and assortative mating. *Journal of Sex Research*, 1974, *10*, 47–51.

Farley, F. H., Mueller, C., Bloomquist. Arousal, personality and assortative mating in marriage: Generability and cross-cultural factors. *Journal of Sex and Marital Therapy*, Spring 1978, *4* (1).

Feld, S. Feelings of adjustment. In F. I. Nye and L. Hoffman (Eds.), *The Employed Mother in America*. Westport, Connecticut: Greenwood Press, 1976, 331–352.

Feldman, H. and Feldman, M. *The Relationship between the Family and Occupational Functioning in a Sample of Rural Women*. Ithaca, New York: Department of Human

Development and Family Studies, Cornell University, 1973.

Feldman, H., Summers, S., and Kiesler, S. B. Those who are number two try harder: The effect of sex on attributions of casuality. *Journal of Personality and Social Psychology*, 1974, *30*, (6) 846–855.

Felson, M. and Knoke, D. Social status and the married woman. *Journal of Marriage and the Family*, August 1974, *36*, 516–521.

Fengler, A. Romantic love in courtship: Divergent paths of male and female students. *Journal of Comparative Family Studies*, 1974, *5*, 134–139.

Figley, C. R. Child density and the marital relationship. *Journal of Marriage and the Family*, May 1973, *35*, 272–282.

Fishbein, M. and Burgess, E. W. *Successful Marriage*. Garden City, New York: Doubleday and Company, 1963.

Fisher, W. A. and Byrne, D. Sex differences in response to erotica: love versus lust. *Journal of Personality and Social Psychology*, 1978, *36*, 117–125.

Flexner, S. B. *I Hear America Talking*. New York: Simon and Schuster (Touchstone Edition) 1979.

Folson, J. K. The romantic complex and cardiac respiratory love, In *The Family*. New York: Wiley, 1934.

Frank, E., Anderson, C. and Rubenstein, D. Marital role ideals and perception of marital role behavior in distressed and non-distressed couples. *Journal of Marital and Family Therapy*, January 1980.

Freiden, A. The U.S. marriage market. In. T. W. Schultz (Ed.). *Economics of the Family: Marriage, Children and Human Capital*. Chicago: The University of Chicago Press, 1974, 352–371.

Freud, S. *Contributions to the Psychology of Love: A Special Type of Object Choice Made by Men (1910)*. In collected papers (Vol. 1), New York: Basic Books, 1959.

Gagnon, J. H. *Human Sexualities*. Dallas, Texas: Scott Foresman, and Co., 1977.

Galton, F. *Hereditary Genius*. New York: Macmillan Co., 1869.

Gettings, F. *The Book of the Hand: An Illustrated History of Palmistry*. London: Hamlyn, 1965.

Gilbert, S. J. Self-disclosure, intimacy and communication in family. *Family Coordinator*, 1976, *25*, 221–231.

Glenn, N. D. The contribution of marriage to the psychological well-being of males and females. *Journal of Marriage and the Family*, August 1975, *37*, 594–600.

Glick, P., and Norton, A. J. Marrying, divorcing and living together in the U.S. today. *Population Bulletin*, 1978, *32*, 3–38.

Glick, P., and Norton, A. J. Marrying, divorcing and living together in the U.S. today, In Loparta, H. (Ed.), Family Factbook, Mariarius Academic Media, Chicago, 1980, 197.

Godwin, J. *The Mating Trade*. Garden City, New York: Doubleday and Company, 1973.

Goldstein, J. W., and Rosenfeld, H. M. Insecurity and preference for persons similar to oneself. *Journal of Personality*, 1969, *37*, 253–268.

Good, L. R., and Good, D. C. On perceiving probability of marital success. *Psychological Reports*, 1972, *31*, 300–302.

Goods, W. J. The theoretical importance of love. *American Sociological Review*, 1959, *24*, 38–47.

Goodwin, M. Expressed self-acceptance and interspousal needs: A basis of mate selection. *Journal of Counseling Psychology*, 1964, *11*(2)

Gove, W. R. Sex, marital status, and suicide. *Journal of Health and Social Behavior*, June 1972, *13*, 204–13.

Gove, W. R. Sex, marital status and morality. *American Journal of Sociology*, July 1973, *79*, 45–67.

Granvold, D. K., Pedler, L. M., and Schellie, S. G. A study of sex role expectancy and female post divorce adjustment. *Journal of Divorce*, Summer 1979, *2*, 383–393.

Gronseth, E. The breadwinner trap, In Louise Kapp Howe (Ed.) *The Future of the Family*. New York: Simon and Schuster, 1972, 175–191.

Gross, R. H. and Arvey, R. D. Marital satisfaction, job satisfaction, and task distribution in the homemaker job. *Journal of Vocational Behavior*, 1977, *11*, 11–13.

Grube, J. W., Greenstein, T. N., Rankin, W. L. and Kearney, K. A. Behavior change following self-confrontation: a test of the value-medication hypothesis. *Journal of Personality and Social Psychology*, 1977, *35*, 212–216.

Grush, J. and Yehl, J. Marital roles, sex differences and interpersonal attraction. *Journal of Personality and Social Psychology*, 1979, *37* (1), 116–123.

Haavio-Mannila, E. Satisfaction with family, work, leisure and live among men and women. *Human Relations*, 1971, *24*, 585–601.

Hannan, M., Tuma, N., and Groeneveld, L. Income and independence effects on marital dissolution: Test of a model. *Paper Presented at American Sociological Association Meetings*, Chicago, August 1977.

Harlow, H. F. *Learning to Love*. San Francisco: Albion, 1971.

Hatfield, E., Utne, M. K., and Traupmann, J. Equity theory and intimate relationships. In R. L Burgess and T. L. Huston (Eds.) *Social Exchange in Developing Relationships*. New York: Academic Press, 1979.

Hatkoff, T. S., and Laswell, T. Love and age, sex and life course experiences. Paper presented at National Council on Family Relations Convention, San Diego, 1977.

Haug, M. Sex role variations in occupational prestige ratings. *Sociological Forces*, January 1975, *8*, 47–56.

Hayes, M. P., Stinnett, N., and Defrain, J. Learning about marriage from the divorced. *Journal of Divorce*. Fall 1980, *4*(1).

Heath, A. *Rational Choice and Social Exchange*. London: Cambridge University Press, 1976.

Herman, S. J. Women and Divorce. *Journal of Divorce*, Winter 1977, *1*(2).

Hetherington, E. M. Effects of father absence on personality development in adolescent daughters. *Developmental Psychology*, 1972, *7*, 313–326.

Hetherington, E. M. Divorce: A child's perspective. *American Psychologist*, October 1979, *34* (10), 851–858.

Hetherington, E., Cox, M., and Cox, R. The aftermath of divorce. In J. H. Stevens, Jr., and M. Matthews (Eds.) *Mother Child, Father Child Relations*. Washington, D.C.: NAEVC, 1978.

Hetherington, E. M., Cox, M., and Cox, R. Family interactions and the social emotional and cognitive development of children following divorce. Paper Presented at the Johnson and Johnson Symposium on the Family: Setting Priorities, Washington, D.C., 1978.

Hill, C., Rubin, Z., and Peplan, L. Breakups before marriage: The end of 103 affairs. *Journal of Social Issues*, Winter 1976, *32*, 147–168.

Hiller, D. V., and Philliber, W. W. Necessity, compatability and status attainment as factors in the labor-force participation of married women. *Journal of Marriage and the Family*, 1980, *42*, 103–110.

Hiller, D. V., and Philliber, W. W. Predicting marital and career success among dual-worker couples. *Journal of Marriage and the Family*, February 1982.

Hinkle, D. E., and Sporakowski, M. J. Attitudes towards love: Reexamination. *Journal of Marriage and the Family*, 1975, *37*, 764–767.

Hobart, C. W. Disillusionment in marriage and romanticism. *Marriage and Family Living*, 1958, *20*, 156–162.

Hobart, C. W., and Lindhom, L. The theory of complementary needs: a reexamination. *Pacific Sociological Review*, 1963, *6*, 73–79.

Hobart, C. W. The incidence of romanticism during courtship. *Social Forces*, 1958, *36*, 362–367.

Holstrom, L. L. *The Two-Career Family*. Cambridge: Shenkman, 1972.

Holter, H. *Sex Roles and Social Structure*. Oslo, Norway: Universite Loraget, 1970.

Homans, G. C. *Social Behavior: Its Elementary Forms*. (Rev. ed.) New York: Harcourt Brace Jovanovich, 1974.

Hoon, P. W., Wincze, J. P, and Hoon, E. F. A test of reciprocal inhibition: Are anxiety and sexual arousal in women mutually inhibitory? *Journal of Abnormal Psychology*, 1977, *86*, 65–74.

Hornung, C. A., and McCullough, B. C. Status and inconsistency in marriage: consequences for life and marital dissatisfaction. Paper Presented at the Annual Meeting of the American Sociological Association, Chicago, 1977.

Horrocks, J. E., and Jackson, D. W. *Self and Role: A Theory of Self-Process and Role Behavior*. Boston: Houghton Mifflin, 1972.

How Long Till Equality? *Time Magazine*, July 12, 1982, p. 23.

Howard, J. and Dawes, R. Linear predictions of marital happiness. *Personality and Social Psychology Bulletin*. Fall 1976, *2*, 478–480.

Hunt, M. *Sexual Behavior in the 1970's*. New York: Dell, 1974.

Huser, W. R. and Grant, C. W. A study of husbands and wives from dual-career and traditional career families. *Psychology of Women Quarterly*, 1978, *31*, 78–79.

Huston, T. L. Ambiguity of acceptance, social desirability, and dating choice. *Journal of Experimental and Social Psychology*, 1973, *9*, 32–42.

Huston, T. L. A decision-making model of the initiation of heterosexual pair relationships. Paper Presented at the National Council in Family Relations, Toronto, October, 1973.

Huston, T. L. A perspective on interpersonal attraction. In T. L. Huston (Ed.), *Foundations of Interpersonal Attraction*. New York: Academic Press, 1974.

Hutton, S. P. Self-esteem, values and self-differentiation in premarital dyads. In B. Murstein (Ed.) *Who Will Marry Whom? Theories and Research in Marital Choice*. New York: Springer Publishing Company, 1976, 237–252.

Jacobson, D. S. The impact of marital separation/divorce on children: I Parent child separation and child adjustment. *Journal of Divorce*, *1*, 341–360.

Jacobson, D. S. The impact of marital separation/divorce on children: II Interparent hostility and child adjustment. *Journal of Divorce*, 1978, *2*, 3–19.

Jacobson, D. S. The impact of marital separation/divorce on children: III Parent child communication and child adjustment, and regression analysis of findings from overall study. *Journal of Divorce*, 1978, *2*, 175–194.

Jacoby, S. Forty-nine million singles can't all be right. *The New York Times Magazine*, February 17, 1974. 12*ff*.

Johnson, R. E. Some correlates of extramarital coitus. *Journal of Marriage and the Family*, August, 1970, 449–454.

Johnson, R. E. Attitudes toward extramarital relationships. *Medical Aspects of Human Sexuality*, 1972, 6(4), 168–191.

Jorgensen, S. R. and Klein, D. M. Sociocultural heterogamy,

dissensus, and conflict in marriage. *Pacific Sociological Review*, 1979, *22*, 31–75.

Jusenius, C. The influence of work experience, skill requirements and occupational segregation on women's earnings. *Journal of Economics and Business*, 1977, *29*, 107–115.

Kalish, R. A. and Knudtson, F. W. Attachment versus disengagement: a life-span conceptualization. *Human Development*, 1976, 171–181.

Kalter, N. Children of divorce in an outpatient psychiatric population. *American Journal of Orthopsychiatry*, 1977, *47* (1), 40–51.

Kanin, E. J., Davidson, D. K. D, and Scheck, S. R. A research note on male-female differentials in the experience of heterosexual love. *Journal of Sex Research*, 1970, *6*, 64–72.

Kanter, R. M. *Men and Women of the Corporation*. New York: Basic Books, 1977.

Kanter, R. M., Jaffee, D., and Weisberg, D. K. Coupling, parenting and the presence of others. Intimate relations in communal households. *The Family Coordinator*, 1975, *24*, 433–452.

Karp, S., Jackson, J. H., and Lester, D. Ideal self-fulfillment in mate selection: A corollary to the complementary need theory of mate selection. *Journal of Marriage and the Family*, 1970, *32*, 269–272.

Kassner, M. W. Will both spouses have careers? Predictions of preferred traditional or egalitarian marriages among university students. *Journal of Vocational Behavior*, 1981, 18, 340–355.

Keith, P. M., and Brubaker, T. H. Sex role expectations associated with specific household tasks: perceived age and employment differences. *Psychological Reports*, 1977, *41*, 15–18.

Kelley, J. B., and Wallerstein, J. S. The effects of parental divorce: Experiences of the child in early latency. *American Journal of Orthopsychiatry*, 1976, *46*, 20–30.

Kemodle, W. Some implications of the homagamy–complementary needs theories of mate selection for sociological research. *Social Forces*, 1969, *38*, 142–152.

Kent, D. P. Subjective factors in mate selection: An exploratory study. *Sociology and Social Research*, 1951, *35*, 391–398.

Kephart, W. M. Evaluation of romantic love. *Medical Aspects of Human Sexuality*, February 1973, 7, 92–112.

Kerckhoff, A. C. Status related value patterns among married couples. *Journal of Marriage and The Family*, February 1972.

Kerckhoff, A. C. Status related value patterns among married couples. *Journal of Marriage and the Family*, February 1972, *34*, 105–110.

Kerckhoff, A. C. The social context of interpersonal attraction. In T. L. Huston (Ed.) *Foundations of Interpersonal Attraction*. New York: Academic Press, 1974.

Kerckhoff, A. C., and Bean, K. E. Value consensus and need complementarity in mate selection. *American Sociological Review*, 1962, *27*, 295–303.

Kierdek, L. A., and Siesky, A. E. Effects of divorce on children: the relationships between parent and child perspectives. *Journal of Divorce*, Winter 1980, *4* (2).

Kiesler, S. B., and Baral, R. L. The search for a romantic partner: the effects of self-esteem and physical attractiveness on romantic behavior in K. Gergen and D. Marlow (Eds.) *Personality and Social Behavior*, Addison-Wesley, 1970, 155–165.

Kindelan, K. and McCarrey, M. Spouses' value systems: similarity, type and attributed marital adjustment. *Psychological Reports*, December 1979, *44*, 1295–1302.

Kinsey, A. C., Pomeroy, W. B., Martin, C. E., and Gebhard, P. H. *Sexual Behavior in the Human Female*. Saunders, Philadelphia, 1953.

Kinsey, A. C., Pomeroy, W. B., and Martin, C. E. *Sexual Behavior in the Human Male*. Saunders, Philadelphia, 1948.

Kipnis, D. Changes in self-concepts in relation to perceptions of others. *Journal of Personality*, 1961, *29*, 449–465.

Kitson, G. C., and Rashke, H. Divorce Research: What we know; what we need to know. *Journal of Divorce*, Spring, 1981, *4*, 1–38.

Kitson, G. C., and Sussman, M. B. Marital complaints, demographic characteristics, and symptoms of mental distress in divorce. *Journal of Marriage and the Family*, February 1982.

Klemesrud, J. A wife's role in big decisions. *N. Y. Times*, November 23, 1980. p. 68.

Knudtson, F. W. Life-span attachment: complexities, questions, considerations. *Human Development*, 1976, *19*, 182–196.

Knupfer, G., Clark, W., and Room, R. The mental health of the unmarried. *American Journal of Psychiatry*, 1966, *122*, 841–851.

Komarovsky, M. Cultural contradictions and sex roles: the masculine case. *American Journal of Sociology*, 1973, *78*, 873–84.

Ktsanses, T. The theory of complementary needs in mate selection: an analytic and descriptive study. *American Sociological Review*, June 1954, 241–249.

Ktsanses, T. Mate selection on the basis of personality type, *American Sociological Review*, 1955, *20*, 547–551.

Kulka, R. A., and Weingarten, H. The long-term effects of parental divorce in childhood on adult adjustment. *Journal of Social Issues*, 1979, *35* (4).

Kurdek, L. A., and Siesky, A. E. Children's perceptions of their parents divorce. *Journal of Divorce*, 1980, *3*, 339–378.

Kurian, G. T. *The Book of World Rankings*. New York: Facts on File, 1979.

Lamb, M. E. The effects of divorce on children's personality development. *Journal of Divorce*, 1977, *1*, 163–174.

Landis, J. T., and Landis, M. G. *Building A Successful Marriage*. Englewood Cliffs, New Jersey: Prentice Hall Inc., 1977.

Landis, J. T. Social correlates of divorce or nondivorce among the unhappily married. *Marriage and Family Living*, 1963, *25*, 178–180.

Landis, J. T., and Landis, M. G. *Building a Successful Marriage*. Englewood Cliffs, New Jersey: Prentice Hall, 1968.

Larzelere, R. E. and Huston, T. L. The dyadic trust scale: toward understanding interpersonal trust in close relationships. *Journal of Marriage and the Family*, August, 1980, *42* (3), 595–604.

Lasswell, T. E. and Lasswell, M. E. I love you but I'm not in love with you. *Journal of Marriage and Family Counseling*, 1976, *2*, 211–224.

Lee, A. and Lee, C. A. *The Total Couple*. New York: Cornerstone Library, 1978.

Lee, J. A. *The Colors of Love*, New York: Bantam Books, 1977.

Lee, J. A. The styles of loving. *Psychology Today*, October 1974, 43–51.

LeMasters, E. E. *Modern Courtships and Marriage*. New York: The Macmillan Company, 1957.

Leslie, G. R. *The Family in the Social Context* (3rd ed.) New York: Oxford University Press, 1976.

Levinger, G. A social psychological perspective on marital dissolution. *The Journal of Social Issues*, 1976, *32* (1), 21–48.

Levinger, G., Senn, D. J., and Jorgensen, B. W. Progress toward permanence in courtship: A test of the Kerckhoff-Davis hypotheses. *Sociometry*, 1970, *33*, 427–433.

Levinger, G. and Snoek, J. D. *Attraction in Relationship: A New Look at Interpersonal Attraction*. New York: General Learning Press, 1972.

Lewis, R. A. A developmental framework for the analysis of premarital dyadic formation. *Family Process*, 1972, *11*, 17–48.

Lewis, R. A. A longitudinal test of a developmental framework for premarital dyadic formation. *Journal of Marriage and the Family*, February 1973.

Libby, R. W., and Whitehurst, R. N. *Marriage and Alternatives: Exploring Intimate Relationships*. Palo Alto, California: Scott Foresman, 1977.

Lindsey, B. and Evans, W. *The Companionate Marriage*. New York: Boni and Liveright, 1927.

Lipetz, M. E., Cohen, I. H., Dworin, J., and Rogers, L. Need complementarity, marital stability and marital satisfaction. In K. J. Gergen and D. Marlowe (Eds.), *Personality and Social Behavior*, Reading, Massachusetts: Addison-Wesley, 1970, 201–212.

Lobsenz, N. M., and Murnstein, B. I. Keeping score: It's fine for football games but it's disastrous for a marriage. *Woman's Day*, September 1976, 146–154.

Locke, H. J. *Predicting Adjustment in Marriage: A Comparison of a Divorced and a Happily Married Group*. New York: Holt, 1951.

Lott, A. J. A possible role for generally adaptive features in mate selection and sexual stimulation. *Psychological Reports*, *45*, 1979.

Lowenthal, M. E., and Weiss, L. Intimacy and crises in adulthood. *The Counseling Psychologist*, 1976, *6*, 88–94.

Luckey, E. B. Marital satisfaction and its concomitant perceptions of self and spouse. *Journal of Counseling Psychology*, 1964, *11*(2).

Luckey, E. B. Number of years married as related to personality perceptions and marital satisfaction. *Journal of Marriage and the Family*, 1966, *28*, 44–48.

Madden, J. F. *The Economics of Sex Discrimination*. Massachusetts: Lexington Books, 1973.

Marini, M. M. Dimensions of marriage happiness: a research note. *Journal of Marriage and the Family*, August 1976, *38*, 443–448.

Markle, G. E. Sex ratio at birth: values variance and some determinants. *Demography*, February 1974, *11*(1).

Markle, G. E., and Nam, C. B. Sex predetermination: its impact on fertility. *Social Biology*, 1971, *18*(1), 73–78.

Maslow, A. H. *Motivation and Personality*. New York: Harper and Row, Publishers, 1970.

Mason, K. O., Czajka, J. L., and Arber, S. Change in U. S. women's sex role attitudes, 1964–1974. *American Sociological Review*, 1976, *41*, 573–596.

Masters, W. H., and Johnson, V. E. *Human Sexual Response*. St. Louis: Brown, Little and Brown, 1966.

Masters, W. H., and Johnson, V. E. *The Pleasure Bond*. New York: Bantam Books, 1975.

Maurois, A. *Disraeli*. New York: Time Reading Program, 1965.

McCary, J. L. *Freedom and Growth in Marriage*. Santa Barbara: Hamilton Publishers, 1975.

McCary, J. L. *McCary's Human Sexuality* (3rd Ed.) New York:

Van Nostrand, 1978.

McDermott, J. Divorce and its psychiatric sequelae in children. *Archives of General Psychiatry* 1970, *23*, 421–428.

McWhirter, N. *Guiness Book of World Records, 1980 Edition*. New York: Sterling Publishing Co., Inc., 1979.

Medling, J. M. and McCarrye, M. Marital adjustment over segments of the family life cycle: the issue of spouses' value similarity. *Journal of Marriage and the Family*, February 1981, 195–203.

Meyer, J. P., and Pepper, S. Need compatability and marital adjustment in young married couples. *Journal of Personality and Social Psychology*, 1977, *35*, 331–342.

Miller, D., Byrne, D., and Denino, J. The relationship between arousal in response to erotic cues and marital adjustment. Unpublished Paper, 1979.

Miller, H. L., and Siegel, P. S. *Loving: A Psychological Approach*. New York: John Wiley and Sons, 1972.

Millett, K. The shame is over. *Ms. Magazine*, January 1975, 26–29.

Molloy, J. T. *Dress for Success*. New York: Peter H. Wyder, 1975.

Morgan, M. *The Total Woman*. New York: Pocket Books, 1975.

Morris, S. *The Book of Strange Facts and Useless Information*. Garden City: Doubleday And Co., Inc. (Dolphin Books) 1979.

Morrison, J. Parental divorce as a factor in childhood psychiatric illness. *Comprehensive Psychiatry*, 1974, *15*, 95–102.

Mueller, C. W., and Pope H. Marital instability: A study of its transmission between generations. *Journal of Marriage and the Family*, 1977, *39*(1), 83–93.

Munro, A. Parent-child separation: Is it really a cause of psychiatric illness in adult life? *Archives of General Psychiatry*, 1969, *20*(5), 598–603.

Murstein, B. I. Physical Attractiveness and marital choice. *Journal of Personality and Social Psychology*, 1972, *22*(1), 8–12.

Murstein, B. I. Empirical Tests of Role, Complementary needs, and homogany theories of marital choice.

Journal of Abnormal and Social Psychology, 1963, *67*, 636–640.

Murstein, B. I. The relationship of mental health to marital choice and courtship progress. *Journal of Marriage and the Family*, 1967.

Murstein, B. I. *Love, Sex, and Marriage Through the Ages*. New York: Springer Publishing Co., 1974.

Murstein, B. I. Empirical Tests of Role, Complementary Needs, and Homogany Theories of Marital Choice. *Journal of Marriage and the Family*, 1967, *29*, 689–696.

Murstein B. I. Stimulus-Value-Role: A theory of marital choice. *Journal of Marriage and the Family*, 1970, *32*, 465–471.

Murstein, B. I. (Ed.) *Theories of Attraction and Love*. New York: Springer, 1971.

Murstein, B. I. Self, ideal-self discrepancy and the choice of marital partner. *Journal of Consulting and Clinical Psychology*, 1971, *37*, 47–52.

Murstein, B. I. A theory of marital choice and its applicability to marriage adjustment and friendship. Pp. 100–151 in B. I. Murstein (Ed.), *Theories of Attraction and Love*. New York: Springer Publishing Company, 1971.

Murstein, B. I. Perceived congruence among premarital couples as a function of neuroticism. *Journal of Abnormal Psychology*, 1973, *81*, 22–26.

Murstein, B. I. Clarification of obfuscation on conjugation: A reply to a criticism of SVR theory of marital choice. *Journal of Marriage and the Family*, 1974, *36*, 231–234.

Murstein, B. I. *Love, Sex and Marriage Through the Ages*. New York: Springer, 1974.

Murstein, B. I. *Who Will Marry Whom? Theories and Research in Marital Choice*. New York: Springer, 1976.

Murstein, B. I. Mate selection in the 1970's. *Journal of Marriage and the Family*, November 1980, 777–792.

Murstein, B. I., and Beck, G. D. Person perception, marriage adjustment and social desirability. *Journal of Consulting and Clinical Psychology*, 1972, *39*, 396–403.

Murstein, B. I., and Christy, P. Physical attractiveness and marriage adjustment in middle-aged couples. *Journal of Personality and Social Psychology*, 1976, *34*, 537–542.

Murstein, B. I., and Glaudin, V. The relationship of marital adjustment to personality: A factor analysis of the interpersonal checklist. *Journal of Marriage and the Family*, 1966, *28*, 37–43.

Neiswender-Reedy, M. E., Birren, J. E., and Schaie, K. W. Love in adulthood: beliefe vs. experience. *Am. Psychol. Assoc. Meeting*, Washington, 1976.

Newcomb, T. M. *The Acquaintance Process*. New York: Holt, Rinehart and Winston, 1961.

New York Magazine, August 16, 1982, 15(32).

Nilson, L. B. The social standing of a married woman. *Social Problems*, June 1976, *23*, 581–592.

Packard, V. *The Sexual Wilderness*. New York: Atheneum, 1981.

Pace, R. *They Went to College*. Minneapolis: University of Minnesota Press, 1941.

Palmer, J., and Byrne, D. Attraction toward dominant and submissive strangers: similarity vs. complementarity. *Journal of Experimental Research in Personality*, 1970, *4*, 108–115.

Pam, A., Plutchik, R., and Conte, H. R. Love: A psychometric approach. *Psychological Reports*, 1975, *37*, 83–88.

Parelius, A. P. Emerging sex-role attitudes. Expectations and strains among college women. *Journal of Marriage and the Family*, 1975, *37*, 146–151.

Parsons, T. Status inequality and stress in marriage. *American Sociological Review*, 1975, *40*, 344–57.

Pearce, D. How long till equality. *Time Magazine*, July 12, 1982.

Pearlin, L. I., and Johnson, J. S. Marital status, life strains, and depression. *American Sociological Review*, 1977, *42*, 704–775.

Peele, S., with Brodsky, A. *Love and Addiction*. New York: Taplinger, 1975.

Pellegrini, R. J., Hicks, R. A., and Meyers-Winston, S. Situational affective arousal and heterosexual attraction: some effects of success, failure and physical attractiveness. *The Psychological Record*, 1979, *29*, 453–462.

Peterson, J. A. *Education for Marriage*. New York: Charles Scribner's Sons, 1964.

Pharis, E. M. and Manosevitz, M. Parental models of infancy: A note on gender preference for firstborns. *Psychological Reports*, 1980, *47*, 763–768.

Pickford, J. H., Signori, E. J., and Rempel, H. Similar or related personality traits as a factor in marital happiness. *Journal of Marriage and the Family*, 1966, *28*, 190–192.

Pickford, J. H., Signori, E. J., and Rempel, H. The intensity of personality traits in relation to marital happiness. *Journal of Marriage and the Family*, 1966, *28*, 458–459.

Pleck, J. H. The male sex role: definitions, problems, and sources of change. *Journal of Social Issues*, 1976, *32*(3), 155–164.

Population Reference Bureau. U. S. women at work. *Population Bulletin*, May 1981, *36*(2).

Prince, A. J. Factors in mate selection. *Family Life Coordinator*, 1961, *10*, 55–58.

Prince, A. J., and Beggaley, A. R. Personality variables and the ideal mate. July–October 1963.

Proxmire, W. Quoted in Time Magazine, March 24, 1975.

Rapoport, R., and Rapoport, R. N. *Dual Career Families*. Baltimore: Penguin Books, 1971.

Rappaport, A. F., and Harrell, J. A behavioral exchange model for marital counseling. *The Family Coordinator*, April 1972, *21*, 203–213.

Reik, T. *A Psychologist Looks at Love*. New York: Holt, Reinhart and Winston, 1944.

Reiter, H. H. Similarities and differences in scores on certain personality scores among engaged couples. *Psychological Reports*, 1970, *26*, 465–466.

Renne, K. S. Correlates of dissatisfaction in marriage. *Journal of Marriage and the Family*, February 1970, *32*, 54–67.

Richardson, J. G. Wife occupational superiority and marital troubles: An examination of the hypothesis, *Journal of Marriage and the Family*, February 1979, *41*, 63–72.

Rim, Y. Sex-typing and means of influence in marriage. *Social Behavior and Personality*, 1980, *8*(1), 117–119.

Rollins, B., and Cannon, K. Marital satisfaction over the family life cycle: Re-evaluation. *Journal of Marriage and the Family*, November 1974, *36*, 771–782.

Rollins, R., and Feldman, M. Marital satisfaction through the life span. *Journal of Marriage and the Family*, 1970, *32*, 20–27.

Rougemont, D. *Love in the Western World*. New York: Pantheon, 1956.

Rubin, Z. Measurement of romantic love. *Journal of Personality and Social Psychology*, 1970, *16*, 265–273.

Rubin, Z., and Levinger, G. Theory and data badly mated: A critique of Murstein's SVR and Lewis' PDF models of mate selection. *Journal of Marriage and the Family*, 1974, *36*, 226–231.

Rubin, Z., Hill, C., Peplau, T., Letita, A. and Dunkel-Schetter, C. Self disclosure in dating couples: sex roles and the ethic of openness. *Journal of Marriage and the Family*, May 1980, 305–317.

Ryder, R. G., Kafka, J. S., and Olson, D. H. Separating and joining influences in courtship and early marriage. *American Journal of Orthopsychiatry*, 1971, *4*, 450–464.

Safilios-Rothschild, C. A Macro- and micro-examination of family power and love: an exchange model. *Journal of Marriage and the Family*, May 1976, *38*, 355–362.

Salamon, S. "Male chauvinism" as a manifestation of love in marriage. *Journal of Asian and African Studies*, 1975, *10*, 1–2.

Sanders, D. *The First of Everything*. New York: Delacorte Press, 1981.

Scanzoni, J. *Sexual Bargaining: Power Politics in the American Marriage*, Englewood Cliffs, N.J.: Prentice-Hall, 1972.

Schaefer, M. T. and Olson, D. H. Assessing intimacy: The Pair Inventory. *Journal of Marital and Family Therapy*, January 1981.

Schellenberg, J. A. Homogany in personal values and the "field of eligibles". *Social Forces*, 1960, *39*, 157–162.

Schmidt, G., Sigusch, V. Sex differences in response to psycho-sexual stimulation by films and slides. *Journal of Sex Research*, 1970, *6*, 268–283.

Schmidt, G., Sigusch, V., and Schafer, S. Response to reading erotic stories: male-female differences. *Archives of Sexual Behavior*, 1973, *2*, 181–199.

Schoen, R. California divorce rates by age at first marriage and duration of first marriage. *Journal of Marriage and the Family*, August 1975, *37*, 548.

Schuman, W. R., Figley, C. R., and Fuhs, N. N. Similarity in self-esteem as a function of duration of marriage among student couples. *Psychological Reports*, 1980, 47, 365–366.

Shanteau, J., Shanteau, N., and Geraldine, F. Probability of acceptance in mating choice. *Journal of Personality and Social Psychology*, April 1979, *37*(4), 522–533.

Shepard, J. W., and Ellis, H. D. Physical attractiveness and selection of marriage partners. *Psychological Reports*, 1972, *30*, 1004.

Shostrom, E. The human encounter. In Otto, *Love Today: A New Exploration*. New York: Association Press, 1972.

Sigusch, V., Schmidt, G., Reinfeld, A., and Wiedmann-Sutor, I. Psychosexual stimulation: sex differences. *Journal of Sex Research*, 1970, *6*, 10–24.

Sihenauer, J., and Carroll, D. *Singles: The New Americans*. New York: Simon and Schuster, 1982.

Silverman, I. Physical attractiveness and courtship. *Sexual Behavior*, September 1971, 22–25.

Sindberg, R. M., Roberts, A. F., and McClain, D. Mate selection factors in computer matched marriages. *Journal of Marriage and the Family*, November 1972, 611–614.

Sorosky, A. D. The psychological effects of divorce on adolescents. *Adolescence*, 1977, *12*(4), 123–136.

Spanier, G., and Cole, C. Toward clarification and investigation of marriage adjustment. *International Journal of Sociology of the Family*, 1976, *6*(1), 121–146.

Staines, G., Pleck, J., Shepard, L., and O'Connor, P. Wives' employment status and marital adjustment: Yet another look. *Psychology of Women Quarterly*, Fall 1978, *3*(1).

Standard and Poor's Industry Surveys, Section 2, May 20, 1982.

Stake, J. E. The ability/performance dimension of self-esteem: Implications for women's achievement behavior, *Psychology of Women Quarterly*, Summer 1979, *3*(4).

Statistical Abstracts of the U. S. National Data Book and Guide to Sources, U.S. Bureau of the Census, 1966–1982.

Stephan, W., Berscheid, E., and Walster, E. Sexual arousal and heterosexual perception. *Journal of Personality and Social Psychology*, 1971, *20*, 93–101.

Stiles, H. R. *Bundling: Its Origin, Progress and Decline in America*. New York: Book Collectors Association Inc. 1934.

Stinnett, N., Carter, L., and Montgomery, J. Older person's perception of their marriages. *Journal of Marriage and the Family*, 1972, *34*, 665–670.

Stinnett, N., and Walters, J. *Relationships in Marriage and Family*, New York: Macmillan, 1977.

Strauss, A. The influence of parent images upon marital choice. *American Sociological Review*, October 1946, *11*(5), 554–559.

Strauss, A. Personality needs and marital choice, *Social Forces*, 1947, *25*, 332–335.

Stroebe, W., Insko, C. A., Thompson, V. D., and Layton, B. P. Effects of physical attractiveness, attitude similarity, and sex on various aspects of interpersonal attraction. *Journal of Personality and Social Psychology*, 1971, *18*, 79–91.

Strong, J. R. A marital conflict resolution model, Redefining conflict to achieve intimacy. *Journal of Marriage and Family Counseling*, 1971, *1*, 269–276.

Sugar, M. Children of divorce. *Pediatrics*, 1970, *46*, 588–595.

Sussman, M. B. Marriage contracts: social and legal consequences. Presented at the 1975 International Workshop on Changing Sex Roles in Family and Society. June 17, 1975.

Teenage Pregnancy. Publication Y3w84: 10116, Washington D.C. U.S. Government Printing Office, 1977.

Teevan, J. J., Jr. Reference groups and premarital sexual behavior. *Journal of Marriage and the Family*, 1972, *34*, 283–291.

Tennov, D. *Love and Limerence*. New York: Stein and Day, 1979.

Terman, L. M. *Psychological Factors in Marital Happiness*. New York: McGraw-Hill.

Tesser, A., and Brodie, M. A note on the evaluation of a computer date. *Psychonamic Science*, 1971, *23*, 300.

The New Scarlet Letter. *Time*, August 2, 1982, 62.

Thornes, B., and Collard, J. *Who Divorces?* London: Routeledge and Kegan Paul, 1979.

Thorton, A. Children and Marital Stability, *Journal of Marriage and the Family*, Aug. 1977, *4*(1).

Toman, W. *Family constellation: Its effects on personality and social behavior* (3rd ed.), New York: Springer Publishing Company, 1976.

Tooley, K. Antisocial behavior and social alienation post divorce: The "man of the house" and his mother. *American Journal of Orthopsychiatry*, 1976, *46*(1), 33.

Touhey, J. C. Sex-role stereotyping and individual differences in liking for the physically attractive. *Social Psychology Quarterly*, 1979, *42*(3), 285–289.

Traupmann, J. *Equity and marriage: a test of equity theory with intimate relationships*. Ph.D. dissertation, University of Wisconsin, 1977.

Treman, D. J., and Terrell, K. Sex and the process of status attainment: a comparison of working women and men. *American Sociological Review*, April 1975, *40*, 174–200.

Troll, L. E., and Smith, J. Attachment through the life span: some questions about dyadic bonds among adults. *Human Development*, 1976, *19*, 156–170.

Twenty facts on women workers. U.S. Department of Labor, Washington, D.C., 1980.

Tyree, A., and Treas, J. The occupational and marital mobility of women. *American Sociological Review*, 1974, *39*, 293–302.

Udry, J. R. The influence of the ideal mate image on mate selection perception. *Journal of Marriage and the Family*, 1965, *27*, 477, 482.

Udry, J. R. Personality match and interpersonal perception as predictors of marriage. *Journal of Marriage and the Family*, 1967, *29*, 722–725.

Ungar, S. The effects of the certainty of self-perception on self-presentation behavior: A test of the strength of self-enhancement motives. *Social Psychology Quarterly*, 1980, *43*(2), 165–172.

U.S. News and World Report, July 5, 1982, p. 10.

Utne, M. K. Equity and intimate relations: a test of the theory in marital interaction. Ph.D. dissertation, University of Wisconsin, 1977.

Verbrugge, L. M. Marital status and health. *Journal of Mar-*

riage and the Family, May 1979, *41*, 267–285.

Veroff, J., Depner, C., Kulka, R., and Douvan, E. Comparison of American motives: 1957 vs. 1976. *Journal of Personality and Social Psychology*, 1980, *39*, 1258–1271.

Veroff, J., Kulka, R. A., and Douvan, E. *Mental Health in America*. New York: Basic Books, 1981.

Waite, L. J. Working wives: 1940–1960. *American Sociological Review*, February 1976, *41*, 65–80.

Walker, K. E. Household work time: its implication for family decisions. *Journal of Home Economics*, 1973, *65*(7), 7–11.

Wallace, I., Wallechinsky, D., Wallace, A., and Wallace, S. *The Books of Lists #2*. New York: William Morrow, 1980. Inc. 1980.

Wallace, K. Factors hindering mate selection. *Sociology and Social Research*, 1960, *44*, 317–325.

Waller, W. *The family: A dynamic interpretation*. New York, Dryden Press, 1951.

Wallerstein, J. S., and Kelly, J. B. The effects of parental divorce: The adolescent experience. In E. J. Anthony and C. Koupernik (Eds.), *The Child in His Family* (vol. 3). New York: Wiley, 1974.

Wallerstein, J. S., and Kelly, J. B. The effects of parental divorce: experiences of the preschool child. *Journal of the American Academy of Child Psychiatry*, 1975, *14*, 600–616.

Wallerstein, J. S., and Kelly, J. B. The effects of parental divorce: Experiences of the child in later latency. *American Journal of Orthopsychiatry*, 1976, *46*, 256–269.

Wallerstein, J. S. Children and parents 18 months after parental separation: Factors related to differential outcome. Paper presented at the National Institute of Mental Health Conference on Divorce. Washington, D. C., February 1978.

Wallerstein, J. S., and Kelly, J. B. California's children of divorce. *Psychology Today*, January 1980.

Walster, E. The effects of self-esteem on romantic liking. *Journal of Experimental Social Psychology*, 1965, *1*, 184–197.

Walster, E. Effect of self-esteem on liking for dates of various social desirabilities. *Journal of Experimental Social Psychology*, 1970, *6*, 248–253.

Walster, E., Aronson, V., and Abrahams, D. Importance of physical attractiveness in dating behavior. *Journal of Personality and Social Psychology*, 1966, *4*(5), 508–516.

Walster, E., Aronson, V., and Roftman, L. Importance of physical attractiveness in dating behavior. *Journal of Personality and Social Psychology*, 1966, *4*, 506–516.

Walster, E., Berscheid, E., and Walster, G. W. New Directions in Equity Research. In L. Berkowitz and E. Walster (Eds.), *Advances of Social Psychology* (Vol 9). New York: Academic Press, 1976.

Walster, E., Traupmann, J., and Walster, G. W. Equity and extramarital sexuality. *Archives of Sexual Behavior*, 1978, *7*, 127–141.

Walster, E., and Walster G. W. *A New Look at Love*. Reading, Massachusetts: Addison Wesley, 1978.

Walster, E., and Walster, G. W. Effect of expecting to be liked on choice of associates. *Journal of Abnormal and Social Psychology*, 1963, *67*, 402–404.

Walster, E., Walster, G. W., and Berscheid, E. *Equity: Theory and Research*. Boston: Allyn and Bacon, 1978.

Walster, E., Walster, W., Piliavin, J., and Schmidt, L. Playing hard-to-get: Understanding an elusive phenomenon. *Journal of Personality and Social Psychology*, 1973, *26*, 113–121.

Walster, E., Walster, G. W., and Traupmann, J. Equity and premarital sex. *Journal of Personality and Social Psychology*, 1978, *37*, 82–92.

Weinraub, M., Brooks, J., and Lewis, M. The social network: a reconsideration of the concept of attachment. *Hum. Dev.*, 1977, *20*, 31–47.

Weiser, M. P. K., and Arbeiter, J. S. *Womanlist*. New York: Atheneum, 1981.

Weiss, R. *Marital Separation*. New York: Basic Books, 1975.

Weiss, R. Single parent households as settings for growing up. Paper presented at the National Institute of Mental Health Conference on Divorce. Washington, D. C., February 1978.

Weitzman, L. J. *The Marriage Contract: Spouses, Lovers and the Law*. New York: Macmillan Publishing Co., Inc. (The Free Press) 1981.

Welch, S. Support among women for the issues of the women's movement. *Sociological Quarterly*, 1975, *16*, 216–227.

Westman, J. The effect of divorce in a child's personality development. *Medical Aspects of Human Sexuality*, 1972, *6*, 38–55.

White, G. L. Physical attractiveness and courtship progress. *Journal of Personality and Social Psychology*, 1980, *39*(4), 660–668.

Whitehouse, J. The role of the initial attracting quality in marriage: Virtues and vices. *Journal of Marital and Family Therapy*, January 1981, 61–65.

Willoughby, R. R. Neuroticism in marriage. *Journal of Social Psychology*, 1963, 7, 19–31.

Wilson, P. P. College women who express futility. New York: Bureau of Publications, Teachers College, Columbia University, 1950, p. 54 (no. 956).

Winch, R. F. Another look at the theory of complementary needs in mate selection. *Journal of Marriage and the Family*, 1967, *29*, 756–767.

Winch, R. F. Some data bearing on the oedipus hypothesis. *Journal of Abnormal and Social Psychology*, 1949, 29, 51–56.

Winch, R. F. *Mate-Selection: A Study of Complementary Needs*. New York: Harper and Row, 1958.

Winch, R. F. The theory of complementary needs in mate-selection: Final results on the test of the general hypothesis. *American Sociological Review*, 1955, *20*, 552–555.

Winch, R. F., Ktsanes, T., and Ktsanes, V. The theory of complementary needs in mate-selection: An analytic and descriptive study. *American Sociological Review*, 1954, *19*, 241–249.

Wolfe, L. *The Cosmo Report*. New York: Arbor House, 1981.

The World Almanac and Book of Facts, 1980, New York: Newspaper Enterprise Association, 1980.

Yohalem, A. M. *The Careers of Professional Women: Commitment and Conflict*. Montclair, N.J.: Allenheld, Osmun Inc., 1978.

Zaleski, Z., and Galkoraska, M. Neuroticism and Marital

Satisfaction. *Behavior Research and Therapy*, 1978, *16*(4), 285–286.

Zuckerman, M. The sensation-seeking motive. In B. Maher (Ed.), *Progress in Experimental Personality Research* (Vol. 7). New York: Academic Press, 1974.

Index

NOTES

NOTES

NOTES